America's Climate Choices

Committee on America's Climate Choices

Board on Atmospheric Sciences and Climate

Division on Earth and Life Studies

NATIONAL RESEARCH COUNCIL
OF THE NATIONAL ACADEMIES

THE NATIONAL ACADEMIES PRESS
Washington, D.C.
www.nap.edu

THE NATIONAL ACADEMIES PRESS · 500 Fifth Street, N.W. · Washington, DC 20001

NOTICE: The project that is the subject of this report was approved by the Governing Board of the National Research Council, whose members are drawn from the councils of the National Academy of Sciences, the National Academy of Engineering, and the Institute of Medicine. The members of the committee responsible for the report were chosen for their special competences and with regard for appropriate balance.

This study was supported by the National Oceanic and Atmospheric Administration under contract number DG133R08CQ0062, Task Order # 4. Opinions, findings, and conclusions, or recommendations expressed in this material are those of the authoring panel and do not necessarily reflect the views of the sponsoring agency.

International Standard Book Number-13: 978-0-309-14585-5 (Book)
International Standard Book Number-10: 0-309-14585-6 (Book)
International Standard Book Number-13: 978-0-309-14586-2 (PDF)
International Standard Book Number-10: 0-309-14586-4 (PDF)
Library of Congress Control Number: 2011927383

Additional copies of this report are available from the National Academies Press, 500 Fifth Street, N.W., Lockbox 285, Washington, DC 20055; (800) 624-6242 or (202) 334-3313 (in the Washington metropolitan area); Internet: http://www.nap.edu

THE NATIONAL ACADEMIES
Advisers to the Nation on Science, Engineering, and Medicine

The **National Academy of Sciences** is a private, nonprofit, self-perpetuating society of distinguished scholars engaged in scientific and engineering research, dedicated to the furtherance of science and technology and to their use for the general welfare. Upon the authority of the charter granted to it by the Congress in 1863, the Academy has a mandate that requires it to advise the federal government on scientific and technical matters. Dr. Ralph J. Cicerone is president of the National Academy of Sciences.

The **National Academy of Engineering** was established in 1964, under the charter of the National Academy of Sciences, as a parallel organization of outstanding engineers. It is autonomous in its administration and in the selection of its members, sharing with the National Academy of Sciences the responsibility for advising the federal government. The National Academy of Engineering also sponsors engineering programs aimed at meeting national needs, encourages education and research, and recognizes the superior achievements of engineers. Dr. Charles M. Vest is president of the National Academy of Engineering.

The **Institute of Medicine** was established in 1970 by the National Academy of Sciences to secure the services of eminent members of appropriate professions in the examination of policy matters pertaining to the health of the public. The Institute acts under the responsibility given to the National Academy of Sciences by its congressional charter to be an adviser to the federal government and, upon its own initiative, to identify issues of medical care, research, and education. Dr. Harvey V. Fineberg is president of the Institute of Medicine.

The **National Research Council** was organized by the National Academy of Sciences in 1916 to associate the broad community of science and technology with the Academy's purposes of furthering knowledge and advising the federal government. Functioning in accordance with general policies determined by the Academy, the Council has become the principal operating agency of both the National Academy of Sciences and the National Academy of Engineering in providing services to the government, the public, and the scientific and engineering communities. The Council is administered jointly by both Academies and the Institute of Medicine. Dr. Ralph J. Cicerone and Dr. Charles M. Vest are chair and vice chair, respectively, of the National Research Council.

www.national-academies.org

COMMITTEE ON AMERICA'S CLIMATE CHOICES

ALBERT CARNESALE (Chair), University of California, Los Angeles
WILLIAM CHAMEIDES (Vice-Chair), Duke University, Durham, North Carolina
DONALD F. BOESCH, University of Maryland Center for Environmental Science, Cambridge
MARILYN A. BROWN, Georgia Institute of Technology, Atlanta
JONATHAN CANNON, University of Virginia, Charlottesville
THOMAS DIETZ, Michigan State University, East Lansing
GEORGE C. EADS, Charles River Associates, Washington, D.C.
ROBERT W. FRI, Resources for the Future, Washington, D.C.
JAMES E. GERINGER, Environmental Systems Research Institute, Cheyenne, Wyoming
DENNIS L. HARTMANN, University of Washington, Seattle
CHARLES O. HOLLIDAY, JR., DuPont (Ret.), Nashville, Tennessee
KATHARINE L. JACOBS,* Arizona Water Institute, Tucson
THOMAS KARL,* NOAA, Asheville, North Carolina
DIANA M. LIVERMAN, University of Arizona, Tucson, and University of Oxford, UK
PAMELA A. MATSON, Stanford University, California
PETER H. RAVEN, Missouri Botanical Garden, St. Louis
RICHARD SCHMALENSEE, Massachusetts Institute of Technology, Cambridge
PHILIP R. SHARP, Resources for the Future, Washington, D.C.
PEGGY M. SHEPARD, WE ACT for Environmental Justice, New York, New York
ROBERT H. SOCOLOW, Princeton University, New Jersey
SUSAN SOLOMON, National Oceanic and Atmospheric Administration, Boulder, Colorado
BJORN STIGSON, World Business Council for Sustainable Development, Geneva, Switzerland
THOMAS J. WILBANKS, Oak Ridge National Laboratory, Tennessee
PETER ZANDAN, Public Strategies, Inc., Austin, Texas

Asterisks (*) denote members who resigned during the course of the study.

NRC Staff:

IAN KRAUCUNAS, Study Director
LAURIE GELLER, Study Director (as of December 2010)
CHRIS ELFRING, Director, Board on Atmospheric Sciences and Climate
PAUL STERN, Director, Committee on the Human Dimensions of Global Change
KATHERINE WELLER, Associate Program Officer
RITA GASKINS, Administrative Coordinator
AMANDA PURCELL, Senior Program Assistant

Foreword:
America's Climate Choices

onvened by the National Research Council in response to a request from Congress (P.L. 110-161), *America's Climate Choices* is a suite of coordinated activities designed to study the serious and sweeping issues associated with global climate change, including the science and technology challenges involved, and provide advice on the most effective steps and most promising strategies that can be taken to respond. The study builds on an extensive foundation of previous and ongoing work, including current and past National Research Council reports, assessments from other national and international organizations, the current scientific literature, climate action plans by various entities, and other sources.

A Summit on America's Climate Choices was convened on March 30–31, 2009, to help frame the study, provide an opportunity for high-level participation and input on key issues, and hear about relevant work carried out by others. Additional outside viewpoints and perspectives were obtained via public events and workshops, invited presentations at meetings, and comments and questions received through the study website *http://americasclimatechoices.org*.

The Panel on Limiting the Magnitude of Future Climate Change was charged to describe, analyze, and assess strategies for reducing the net future human influence on climate, including both technology and policy options. The panel's report focuses on actions to reduce domestic greenhouse gas emissions and other human drivers of climate change, such as changes in land use, but also considers the international dimensions of limiting the magnitude of climate change.

The Panel on Adapting to the Impacts of Climate Change was charged to describe, analyze, and assess actions and strategies to reduce vulnerability, increase adaptive capacity, improve resilience, and promote successful adaptation to climate change in different regions, sectors, systems, and populations. The panel's report draws on a wide range of sources and case studies to identify lessons learned from past experiences, promising current approaches, and potential new directions.

The Panel on Advancing the Science of Climate Change was charged to provide a concise overview of current understanding of past, present, and future climate change, including its causes and its impacts, then recommend steps to advance our current understanding, including new observations, research programs, next-generation models, and the physical and human assets needed to support these and other activities. The panel's report focuses on the scientific advances needed both to improve our understanding of the intergrated human-climate system and to devise more effective responses to climate change.

The Panel on Informing Effective Decisions and Actions Related to Climate Change was charged to describe and assess different activities, products, strategies, and tools for informing decision makers about climate change and helping them plan and execute effective, integrated responses. The panel's report describes the different types of climate change-related decisions and actions being taken at various levels and in different sectors and regions; and it develops a framework, tools, and practical advice for ensuring that the best available technical knowledge about climate change is used to inform these decisions and actions.

The Committee on America's Climate Choices was responsible for providing overall direction, coordination, and integration of the *America's Climate Choices* suite of activities and ensuring that these activities provide well-supported, action-oriented, and useful advice to the nation. The Committee was also charged with writing a final report—this document—that builds on the four panel reports and other sources to answer the following four overarching questions:

- What short-term actions can be taken to respond effectively to climate change?
- What promising long-term strategies, investments, and opportunities could be pursued to respond to climate change?
- What are the major scientific and technological advances needed to better understand and respond to climate change?
- What are the major impediments (e.g., practical, institutional, economic, ethical, intergenerational) to responding effectively to climate change, and what can be done to overcome these impediments?

Collectively, the *America's Climate Choices* suite of activities involved more than 90 volunteers from a range of communities including academia, various levels of government, business and industry, other nongovernmental organizations, and the international community. Study participants were charged to write consensus reports that provide broad, action-oriented, and authoritative analyses to inform and guide responses to climate change across the nation. Responsibility for the final content of each report rests solely with the authoring group and the National Research Council. However, the development of each report included input from and interactions with members of all five study groups; the membership of each group is listed in Appendix A.

Preface

How should the United States respond to the challenges posed by climate change? This is the fundamental question addressed by *America's Climate Choices*—a suite of activities requested by the U.S. Congress and conducted by the U.S. National Research Council. Book shelves and the internet are replete with studies of climate change: Why conduct another one? First among the reasons to do so is that the body of scientific knowledge about climate change is growing rapidly and, as it does, so too does our understanding of the nature and severity of potential consequences. Second, unlike most previous studies, this study looks across the full range of response options and the interactions among them. Third, this work goes beyond analysis of the problem and, in accordance with its Statement of Task, provides "action-oriented advice on what can be done to respond most effectively to climate change..." Toward that end, the committee membership was not limited to physical and social scientists but also included people with expertise and experience in public policy, government, and the private sector.

Numerous substantive and procedural questions arose in the course of the committee's work—for instance, regarding the primary audience to which the final report would be directed. The Statement of Task calls upon the committee to "advise the nation," which indicates an extremely broad audience. Ultimately, the committee chose to view as its audience decision makers at all levels who will influence America's response to climate change. Hence this report's focus on formulating decisions to be made and on strategies for making them. Although this study is focused on America's climate choices and is accordingly directed to American decision makers, the committee's analyses and advice were formulated with full consideration of the international context within which U.S. responses to climate change must be selected and implemented. Another consideration was the analytical framework to use in identifying America's climate choices. Although no single option was selected a priori, the panels and the committee all concluded that iterative risk management is the most useful framework for dealing with the many complexities and uncertainties that are inherent to climate change.

A final example of an issue that required resolution by the committee stems from the assigned task to "provide targeted, action-oriented advice." Some natural and social scientists believe their appropriate role is to provide the best available scientific information, to formulate options for decision makers, and to describe the relative advantages and disadvantages of each of the options. In the views of these individuals, recommending a particular option would carry them beyond objectivity and into advocacy. Others consider it appropriate to inform decision makers of their considered judgments, properly labeled as such. This issue was not resolved in the abstract; rather, the members of the committee sought to achieve consensus on a case-by-case basis. We do recommend specific courses of action where there is substantial evidence supporting the need for such actions, but this advice is fairly general in nature, in a deliberate effort to avoid being "policy prescriptive." Recommendations that deal with government function, such as responsibilities to be assigned to specific federal agencies, were deemed to be beyond the scope of the committee.

Since the time that the Committee began its work, the economic and political context in which climate change decisions are being made has changed a great deal, both domestically and internationally. Within the United States, Congress has considered several substantive proposals for federal legislation related to climate change, but none has become law. The committee did not attempt to analyze these specific proposals or to weigh in with views on other specific political developments taking place during the course of the study.

We hope that the efforts of the panels and this committee will prove useful to the nation as it confronts the complex challenges of climate change in the near term and in the decades ahead. We wish to thank numerous people who provided valuable input to this study, including the following people who were invited guest speakers at the committee's meetings: Anthony Janetos, Joint Global Change Research Institute; Steven Seidel, Harvard University; Jonathan Pershing, U.S. Department of State; Anand Patwardhan, Indian Institute of Technology-Bombay; Richard Suttmeier, University of Oregon; Nicole DeWandre, European Commission; Rik Leemans, Wageningen University; Yvo de Boer, UNFCCC; Franklin Moore, USAID; Ian Noble, World Bank; Scott Barrett, Columbia University; Michael Grubb, U.K. Carbon Trust; Glenn Prickett, Conservation International; Stephen Gardiner, University of Washington; Steven Vanderheiden, University of Colorado; Manuel Pastor, University of Southern California; and Michel Gelobter, Cooler, Inc. Special thanks to Gary Yohe (Wesleyan University; member of the ACC Panel on Adapting to the Impacts of Climate Change) for substantial contributions to the committee's discussions about the concept of risk management. Numerous additional people provided input through participation in the America's Climate Choices Summit and the Geoengineering workshop (see Appendix D for Summit agenda.).

Essential contributions to this project were made by knowledgeable, skilled, and accommodating members of the National Research Council staff, and we are deeply grateful to them. Ian Kraucunas and Laurie Geller were invaluable in organizing and marshalling the effort and in their substantive engagement. We benefitted immensely from the active participation of other members of the staff, especially the important contributions from Chris Elfring, Paul Stern, and Marlene Kaplan, and the outstanding administrative support from Rita Gaskins and Amanda Purcell. Our gratitude extends also to the members of the ACC panels and to the many others who shared with us the knowledge, perspectives, and wisdom essential to the success of America's Climate Choices.

Albert Carnesale (Chair) and William Chameides (Vice Chair)
Committee on America's Climate Choices

Acknowledgments

This report has been reviewed in draft form by individuals chosen for their diverse perspectives and technical expertise, in accordance with procedures approved by the NRC's Report Review Committee. The purpose of this independent review is to provide candid and critical comments that will assist the institution in making its published report as sound as possible and to ensure that the report meets institutional standards for objectivity, evidence, and responsiveness to the study charge. The review comments and draft manuscript remain confidential to protect the integrity of the deliberative process. We wish to thank the following individuals for their participation in their review of this report:

IAN BURTON, Meteorological Service of Canada, Downsview, Ontario
KEN CALDEIRA, Carnegie Institution, Stanford, California
MARTIN J. CHAVEZ, ICLEI Local Governments for Sustainability, Washington, D.C.
WILLIAM C. CLARK, Harvard University, Cambridge, Massachusetts
ROBERT E. DICKINSON, University of Texas, Austin
KIRSTIN DOW, University of South Carolina, Columbia
DAVID GOLDSTON, Natural Resources Defense Council, Washington, D.C.
GEORGE M. HORNBERGER, Vanderbilt University, Nashville, Tennessee
CHARLES KOLSTAD, University of California, Santa Barbara
M. GRANGER MORGAN, Carnegie Mellon University, Pittsburgh, Pennsylvania
RICHARD H. MOSS, University of Maryland, College Park
LAWRENCE T. PAPAY, PQR, LLC, La Jolla, California
SUSAN F. TIERNEY, Analysis Group, Boston, Massachusetts

Although the reviewers listed above have provided many constructive comments and suggestions, they were not asked to endorse the conclusions or recommendations nor did they see the final draft of the report before its release. The review of this report was overseen by **Robert Frosch** (Harvard University) and **Susan Hansen** (Clark University), appointed by the Division on Earth and Life Studies and the Report Review Committee, who were responsible for making certain that an independent examination of this report was carried out in accordance with institutional procedures and that all review comments were carefully considered. Responsibility for the final content of this report rests entirely with the authoring committee and the institution.

Institutional oversight for this project was provided by:

Contents

Summary

Climate change is occurring, is very likely caused by human activities, and poses significant risks for a broad range of human and natural systems. Each additional ton of greenhouse gases emitted commits us to further change and greater risks. In the judgment of the Committee on America's Climate Choices, the environmental, economic, and humanitarian risks of climate change indicate a pressing need for substantial action to limit the magnitude of climate change and to prepare to adapt to its impacts.

This report, the final volume of the *America's Climate Choices* (ACC) suite of activities, examines the nation's options for responding to the risks posed by climate change. Although it is crucial to recognize that climate change is inherently an *international* concern that requires response efforts from all countries, this report focuses on the essential elements of an effective *national* response, which includes:

- Enacting policies and programs that reduce risk by limiting the causes of climate change and reducing vulnerability to its impacts;
- Investing in research and development efforts that increase knowledge and improve the number and effectiveness of response options available;
- Developing institutions and processes that ensure pertinent information is collected and that link scientific and technical analysis with public deliberation and decision making;
- Periodically evaluating how response efforts are progressing and updating response goals and strategies in light of new information and understanding.

Given the inherent complexities of the climate system, and the many social, economic, technological, and other factors that affect the climate system, we can expect always to be learning more and to be facing uncertainties regarding future risks. This is not, however, a reason for inaction. Rather, the challenge for society is to acknowledge these uncertainties and respond accordingly, just as is done in many areas of life. For example, people buy home insurance to protect against potential losses, and businesses plan contingently for a range of possible future economic conditions.

Just as in these other areas, a valuable framework for making decisions about America's Climate Choices is **iterative risk management.** This refers to an ongoing process of identifying risks and response options, advancing a portfolio of actions that emphasize risk reduction and are robust across a range of possible futures, and revising responses over time to take advantage of new knowledge. Iterative risk management strategies must be durable enough to promote sustained progress and long-term investments, yet sufficiently flexible to take advantage of improvements in knowledge, tools, and technologies over time.

In the context of an iterative risk management framework, and building on the analyses in the four ACC panel reports, the committee recommends the following priority actions for an effective and comprehensive national response to climate change:

Substantially reduce greenhouse gas emissions. In the committee's judgment there are many reasons why it is imprudent to delay actions that at least begin the process of substantially reducing emissions. For instance:

- The faster emissions are reduced, the lower the risks posed by climate change. Delays in reducing emissions could commit the planet to a wide range of adverse impacts, especially if the sensitivity of the climate to greenhouse gases is on the higher end of the estimated range.
- Waiting for unacceptable impacts to occur before taking action is imprudent because the effects of greenhouse gas emissions do not fully manifest themselves for decades and, once manifested, many of these changes will persist for hundreds or even thousands of years.
- The sooner that serious efforts to reduce greenhouse gas emissions proceed, the less pressure there will be to make steeper (and thus likely more expensive) emission reductions later.
- The United States and the rest of the world are currently making major investments in new energy infrastructure that will largely determine the trajectory of emissions for decades to come. Getting the relevant incentives and policies in place as soon as possible will provide crucial guidance for these investment decisions.
- In the committee's judgment, the risks associated with doing business as usual are a much greater concern than the risks associated with engaging in strong response efforts. This is because many aspects of an "overly ambitious" policy response could be reversed if needed, through subsequent policy change; whereas adverse changes in the climate system are much more difficult (indeed, on the timescale of our lifetimes, may be impossible) to "undo."

RECOMMENDATION 1: In order to minimize the risks of climate change and its adverse impacts, the nation should reduce greenhouse gas emissions substantially over the coming decades. The exact magnitude and speed of emissions reduction depends on societal judgments about how much risk is acceptable. However, given the inertia of the energy system and long lifetime associated with most infrastructure for energy production and use, it is the committee's judgment that the most effective strategy is to begin ramping down emissions as soon as possible.

Emission reductions can be achieved in part through expanding current local, state, and regional-level efforts, but analyses suggest that the best way to amplify and accelerate such efforts, and to minimize overall costs (for any given national emissions reduction target), is with a comprehensive, nationally uniform, increasing price on CO_2[1] emissions, with a price trajectory sufficient to drive major investments in energy efficiency and low-carbon technologies. In addition, strategically-targeted complementary policies are needed to ensure progress in key areas of opportunity where market failures and institutional barriers can limit the effectiveness of a carbon pricing system.

Begin mobilizing now for adaptation. Aggressive emissions reductions would reduce the need for adaptation, but not eliminate it. Climate change is already happening, and additional changes can be expected for all plausible scenarios of future greenhouse gas emissions. Prudent risk management demands advanced planning to deal with possible adverse outcomes—known and unknown—by increasing the nation's resilience to both gradual changes and the possibility of abrupt disaster events. Effective adaptation will require the development of new tools and institutions to manage climate-related risks across a broad range of sectors and spatial scales. Adaptation decisions will be made and implemented by actors in state and local governments, the private sector, and society at large, but there is also a need for national-level efforts—for instance, to share information and technical resources for evaluating vulnerability and adaptation options, and to develop and implement adaptation plans within the federal agencies and their relevant programs.

RECOMMENDATION 2: Adaptation planning and implementation should be initiated at all levels of society. The federal government, in collaboration with other levels of government and with other stakeholders, should immediately undertake the development of a national adaptation strategy and build durable institutions to implement that strategy and improve it over time.

Invest in science, technology, and information systems. Scientific research and technology development can expand the range, and improve the effectiveness of, options to respond to climate change. Systems for collecting and sharing information, including formal and informal education systems, can help ensure that climate-related decisions are informed by the best available knowledge and analysis, and can help us evaluate the effectiveness of actions taken. Many actors are involved in such efforts. For instance, technological innovation will depend in large part on private sector efforts, while information, education, and stakeholder engagement systems can be advanced by nongovernmental organizations and state and local governments. But the federal government has important roles to play in all of these efforts as well.

RECOMMENDATION 3: The federal government should maintain an integrated, coordinated, and expanded portfolio of research programs with the dual aims of increasing our understanding of the causes and consequences of climate change and enhancing our ability to limit climate change and to adapt to its impacts.

RECOMMENDATION 4: The federal government should lead in developing, supporting, and coordinating the information systems needed to inform and evaluate America's climate choices, to ensure legitimacy and access to climate services, greenhouse gas accounting systems, and educational information.

RECOMMENDATION 5: The nation's climate change response efforts should include broad-based deliberative processes for assuring public and private-sector engagement with scientific analyses, and with the development, implementation, and periodic review of public policies.

Actively engage in international climate change response efforts. America's climate choices affect and are affected by the choices made throughout the world. U.S. emissions reductions alone will not be adequate to avert dangerous climate change risks, but strong U.S. emission reduction efforts will enhance our ability to influence other countries to do the same. Also, the United States can be greatly affected by impacts of climate change occurring elsewhere in the world, and it is in our interest to help enhance the adaptive capacity of other nations. Effectively addressing climate change requires both contributing to and learning from other countries' efforts.

RECOMMENDATION 6: The United States should actively engage in international-level climate change response efforts: to reduce greenhouse gas emissions through cooperative technology development and sharing of expertise, to enhance adaptive capabilities (particularly among developing nations that lack the needed resources), and to advance the research and observations necessary to better understand the causes and effects of climate change.

Coordinate national response efforts. Individuals, businesses, state and local governments, and other decision makers nationwide are already taking steps to respond to climate change risks, and must continue to play essential roles in our nation's future response strategies. Numerous federal government agencies and organizations must also be involved in informing and implementing America's climate choices. Our nation needs a coherent strategy for assuring adequate coordination among this wide array of actors. This includes, for instance, carefully balancing rights and responsibilities among different levels of government (vertical coordination), assuring effective delineation of roles among different federal agencies (horizontal coordination), and promoting effective integration among the different components of a comprehensive climate change response strategy (e.g., all the various efforts discussed in each of the previous recommendations).

RECOMMENDATION 7: The federal government should facilitate coordination of the many interrelated components of America's response to climate change with a process that identifies the most critical coordination issues and recommends concrete steps for how to address these issues.

Responding to the risks of climate change is one of the most important challenges facing the United States and the world today and for decades to come. America's climate choices will involve political and value judgments by decision makers at all levels. These choices, however, must be informed by sound scientific analyses. This report recommends a diversified portfolio of actions, combined with a concerted effort to learn from experience as those actions proceed, to lay the foundation for sound decision making today and expand the set of options available to decision makers in the future.

CHAPTER ONE

The Context for America's Climate Choices

The United States lacks an overarching national strategy to respond to climate change.

America's response to climate change is ultimately about making choices in the face of risks: choosing, for example, how, how much, and when to reduce greenhouse gas (GHG) emissions and to increase the resilience of human and natural systems to climate change. These choices will in turn influence the rate and magnitude of future climate change and its impact on people and many things that people care about. Each course of action carries potential benefits and risks, only some of which can be fully anticipated and quantified. A key question surrounding America's climate choices is thus how we as a society perceive, evaluate, and respond to risk. This question is complicated by the diversity of people, communities, and interests affected by climate change (and by many of the proposed responses to climate change), by their different perceptions and judgments of and tolerances for risk, and by the fact that climate change is an issue that spans local to global scales and multiple generations.

This report, the final volume of the *America's Climate Choices* (ACC) suite of activities (Box 1.1), offers advice on how to weigh the potential risks and benefits associated with different actions that might be taken to respond to climate change, and how to ensure that actions are as effective as possible. America's climate choices will ultimately be made by elected officials, business leaders, individual households, and other decision makers across the nation; and these choices almost always involve tradeoffs, value judgments, and other issues that reach beyond science. The goal of this report, and of the entire ACC suite of activities, is to ensure that the nation's climate choices are informed by the best possible scientific knowledge and analysis, both now and in the decades ahead.

This chapter briefly reviews some key elements of the current context in which America's climate change must be made. Chapter 2 reviews current scientific understanding of the causes and consequences of climate change. Chapter 3 describes some of the features of climate change that make it such a unique, challenging issue to address. Chapter 4 explores the concept of iterative risk management as an over-

---

BOX 1.1
The America's Climate Choices Study

In 2008, Congress commissioned the National Academy of Sciences to: "*investigate and study the serious and sweeping issues relating to global climate change and make recommendations regarding what steps must be taken and what strategies must be adopted in response to global climate change, including the science and technology challenges thereof.*" In response to this mandate, and with financial support from the National Oceanic and Atmospheric Administration, the *America's Climate Choices* suite of activities was established. A Summit on America's Climate Choices held in March 2009, and independent panels were convened to study and produce reports focusing on four specific aspects of responding to climate change: *Advancing the Science of Climate Change,*[a] *Limiting the Magnitude of Future Climate Change,*[b] *Adapting to the Impacts of Climate Change,*[c] and *Informing an Effective Response to Climate Change.*[d]

The panel reports offer a detailed analysis of possible actions and investments in each of these four realms—for instance, regarding specific technologies and policies for reducing greenhouse gas emissions, needs and opportunities for adapting to climate change impacts, and key research needs in different areas of climate change science (see Appendix C for more details about the content of the panel reports). This final report by the Committee on America's Climate Choices draws on the information and analysis in the four panel reports, as well as a variety of other sources, to identify cross-cutting challenges and offer both a general framework and specific recommendations for establishing an effective national response to climate change.

---

[a] National Research Council (NRC), *Advancing the Science of Climate Change* (Washington, D.C.: National Academies Press, 2010).
[b] National Research Council (NRC), *Limiting the Magnitude of Future Climate Change* (Washington, D.C.: National Academies Press, 2010).
[c] National Research Council (NRC), *Adapting to the Impacts of Climate Change* (Washington, D.C.: National Academies Press, 2010).
[d] National Research Council (NRC), *Informing an Effective Response to Climate Change* (Washington, D.C.: National Academies Press, 2010).

arching framework for responding to climate change. Chapter 5 offers recommendations for the key elements of an effective, robust U.S. response.

GREENHOUSE GAS EMISSION TRENDS

Despite an international agreement signed by the United States and 153 other nations in 1992 to stabilize atmospheric GHG concentrations "at a level that would prevent dangerous anthropogenic interference with the climate system,"[1] GHG emissions have continued to rise. Energy-related carbon dioxide (CO_2) emissions constitutes roughly

---

BOX 1.2

Non-CO$_2$ Greenhouse Gases and Aerosols

International negotiations and domestic policy debates have focused largely on reducing CO$_2$ emissions from fossil fuels, both because these emissions account for a large fraction of total GHG emissions and because they can be estimated fairly accurately based on fuel-use data.[a] This report follows suit by focusing primarily on energy-related CO$_2$ emissions. It is important to recognize, however, that there are other important sources of CO$_2$ (such as tropical deforestation), and there are other compounds in the atmosphere that affect the earth's radiative balance and thus play a role in climate change.

This includes long-lived GHGs such as methane, nitrous oxide, and fluorinated compounds (which arise from a variety of human activities including agriculture and industrial activities). It also includes shorter-lived gases that are precursors to tropospheric ozone (which directly affects human health, in addition to influencing climate), and a variety of aerosols that can exert either warming or cooling effects, depending on their chemical and physical properties. Some of these other compounds are explicitly included in climate policy negotiations and emissions reductions plans, but in general it is much more difficult to measure and verify reductions in emissions of many of these substances than for CO$_2$.[b] See NRC, *Advancing the Science of Climate Change* and *Limiting the Magnitude of Climate Change* for more extensive discussion of these other gases and aerosols.

---

[a] It should be noted, however, that there is as yet no sufficiently accurate way to verify countries' self-reported estimates using independent data. A recent study (NRC, *Verifying Greenhouse Gas Emissions: Methods to Support International Climate Agreements*, Washington, D.C.: National Academies Press, 2010) recommended a set of strategic investments that would improve self-reporting and provide a verification capability within 5 years.

[b] NRC, *Verifying GHG Emissions*.

---

83 percent of total U.S. GHG emissions[2] and thus will be the primary focus of the discussion in this report (non-CO$_2$ GHGs are discussed briefly in Box 1.2). Figure 1.1 shows recent and projected energy-related CO$_2$ emissions for the United States. The increase in emissions over the past few decades occurred despite the fact that the "intensity" of America's CO$_2$ emissions (the amount of emissions created per unit of economic output, often presented as emissions per dollar of GDP) decreased by almost 30 percent.[3] Thus, the general tendency for industrialized nations to become more efficient and less carbon intensive has slowed but not prevented the growth of domestic CO$_2$ emissions.

The upward trend in U.S. emissions has been punctuated by brief declines, usually during economic downturns. By far the most significant of these downturns was the roughly 6 percent decrease in energy-related CO$_2$ emissions in 2009, related to the economic recession.[4] However, the U.S. Energy Information Administration's latest

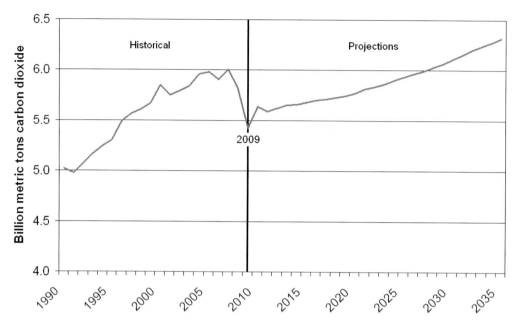

FIGURE 1.1 Energy-related U.S. CO_2 emissions for 1980-2009 (estimated) and 2010-2035 (projected). Given in billion metric tons CO_2. The long-term upward trend in emissions has been punctuated by declines during economic downturns, most notably around 2009. SOURCE: Adapted from Energy Information Administration (EIA), *International Energy Outlook*, Report # DOE/EIA-0484(2010) (Washington, D.C.: U.S. Department of Energy, 2010, available at *http://www.eia.doe.gov/oiaf/ieo/*, accessed March 4, 2011).

projections are that emissions return to an upward trend in 2010, and that under a "business-as-usual" scenario through 2035 (that is, assuming no major actions to reduce domestic GHG emissions or additional major economic downturns), CO_2 emissions will grow by an average of roughly 0.2 percent per year.[5] This is a slower growth rate than that of the past three decades, but it does indicate that emissions will exceed pre-recession levels by the year 2028, and by the year 2035 they will be about 8.5 percent higher than pre-recession levels.

Recent studies have highlighted the commitments to further climate change that are implied by construction of new, long-lived infrastructure (e.g., electricity production facilities, highways). Once constructed, the emissions from these facilities can be locked in for as much as 50 years or more. This is an especially serious concern in regards to rapidly developing countries, where huge investments in new energy generation and energy use systems are being made.[6] As a result, global GHG emissions are projected to increase steeply. The U.S. Energy Information Administration estimates that by 2035, global emissions will be more than 40 percent larger than in 2007 in the absence of aggressive policies to reduce emissions, with most of the increase expected to occur in developing economies.[7]

THE CURRENT CONTEXT

As a signatory to the Copenhagen Accord in 2009, the United States has endorsed an effort to work with the international community to prevent a 2°C (3.6°F) increase in global temperatures relative to pre-industrial levels (see Box 1.3).[8] As part of this accord, the Obama Administration set a "provisional" target of reducing U.S. GHG emis-

---

BOX 1.3
International Context

The ACC studies focus primarily on domestic action, but climate change is an inherently global problem, and U.S. response strategies must be formulated in the context of international agreements and the actions of other nations. The United Nations Framework Convention on Climate Change (UNFCCC) has thus far been the most visible arena for international negotiations on climate change. The only major binding agreement to emerge from the UNFCCC process— the Kyoto protocol—was never ratified by the United States and will expire in 2012. There is no comprehensive agreement for governing response to climate change at the international level. Instead, we have seen the emergence of a loosely coupled "complex" of activities with no clear core, which includes, for instance, bi-lateral initiatives (e.g., the U.S.-China Partnership on Climate Change), clubs of countries that pledge cooperative efforts (e.g., G-8+5 climate change dialogue, Major Economies Forum, Asia Pacific Partnership), and programs of specialized UN agencies (WMO, UNEP, UNDP, FAO) and other international entities (GATT, WTO, World Bank).[a]

Climate change science is also an inherently international enterprise that has been greatly advanced through programs that coordinate and facilitate cooperative multi-national research efforts, such as the World Climate Research Program. Global observing systems, which provide crucial information about climate system variability and long term change, are advanced through cooperative efforts such as the Global Earth Observation System of Systems (GOESS). Also of great importance is the work of the Intergovernmental Panel on Climate Change (IPCC), which synthesizes and translates research developments into information that is useful for policy makers. The United States has been a major contributor to, and beneficiary of, all of these research, observational, and assessment activities.

---

[a] R. O. Keohane and D. G. Victor, *The Regime Complex for Climate Change*. Discussion Paper 2010-33. (Cambridge, Massachusetts: Harvard Project on International Climate Agreements, 2010).

sions in the range of 17 percent below 2005 levels by the year 2020. Given the GHG emission projections discussed in the preceding section, it is clear that the United States will not be able to meet such a commitment without a significant departure from "business-as-usual."

The federal government has adopted some policies (such as subsidies and tax credits) to catalyze the development and implementation of climate-friendly technologies, and there are also a range of voluntary federal programs in place to encourage energy efficiency and GHG emission reductions. More comprehensive federal-level legislation remains stalled. In June 2009, the House of Representatives passed the American Security and Clean Energy Act, which would have established a cap-and-trade system designed to lower U.S. GHG emissions by 17 percent by 2020 and 80 percent by 2050. A similar bill failed to reach the Senate floor, however; and following the 2010 midterm elections, the prospects of any significant climate legislation being passed in the near future have diminished further.

In 2007 the U.S. Supreme Court instructed the U.S. Environmental Protection Agency (EPA) that it is required under the Clean Air Act to regulate emissions of CO_2 and five other greenhouse gases if it finds that such emissions threaten the public health and welfare.[9] In 2009 the EPA issued such a finding and, as a consequence, the EPA is currently developing regulations on GHG emissions from newly constructed or modified power plants and industrial sources;[10] recently, together with the National Highway Traffic Safety Administration, it issued a coordinated set of fuel economy and GHG emissions standards for light-duty vehicles. However, there are many obstacles in the path to EPA regulation (including, for example, potential congressional legislation that would delay or rescind EPA's authority, and litigation likely to follow rulemaking efforts that would use the judiciary to do the same). Thus the timing and character of regulatory programs to control GHG emissions are by no means certain.

Despite the current lack of comprehensive national policies, early actors at other levels of government and in the private sector are advancing policies and commitments to reduce emissions and lessen impacts. In the private sector, many corporations have made commitments or developed action plans for significantly reducing emissions from their operations.[11] More than 1,000 mayors have signed onto the U.S. Conference of Mayors' Climate Protection Agreement, pledging to reduce their city's overall emissions by 7 percent below 1990 levels by 2012.[12] A majority of states have adopted some form of renewable portfolio standard,[13] energy efficiency program requirements, or emissions reduction goal, and some have adopted or plan to adopt cap and trade systems to reduce GHG emissions (for example, the Regional Greenhouse Gas Initiative of the northeastern U.S. states, the Western Climate Initiative, the Midwestern Greenhouse Gas Reduction Accord). The California Global Warming Solutions Act (AB32) enacted a sweeping set of GHG emission control programs for the state (and it survived a ballot proposition to repeal the Act in 2010).

Many states and communities have also developed policies to expand mass transit systems, discourage urban sprawl, increase efficiency, and tighten the energy provisions of building codes,[14] and there are important developments being led by subnational governments and nongovernment organizations in the development of protocols and registries for reporting and verifying GHG missions (e.g., the Climate Registry, the California Climate Action Registry, the Carbon Disclosure Project of ICLEI—Local Governments for Sustainability).[15]

Climate change adaptation planning efforts are also under way in a number of states, counties, and local communities.[16] Adaptation strategies are being explored in climate-sensitive sectors such as agriculture and water resources management, which have historically adapted to natural climate variability in ways that may reduce vulnerabilities to climate change. Several non-governmental organizations have also become active in promoting adaptation planning. In 2009, the White House Council on Environmental Quality, the Office of Science and Technology Policy, and the National Oceanic and Atmospheric Administration initiated an Interagency Climate Change Adaptation Task Force to recommend adaptation initiatives both domestically and internationally. The U.S. intelligence community is assessing how climate change may affect national security (for instance, through geopolitical destabilization from water scarcity or sea level rise), and the U.S. military has begun to consider how climate change will affect their facilities, capabilities, and theatres of operation.[17]

The collective effect of these local, state, federal, and private sector efforts to limit and adapt to climate change is potentially quite significant but, as suggested by recent analyses, it is not likely to yield emission reductions comparable to what could be achieved with strong federal policies.[18] Moreover, it is not clear if the current patchwork of initiatives will prove durable in the absence of an overarching federal policy. For example, evidence suggests that many early actors have been motivated at least in part by a belief that federal legislation on climate change is inevitable and that getting out in front of that legislation will offer a competitive advantage.[19] Without a federal policy, emission cuts made in states with climate programs may be undermined by "leakage" to states without such programs, and varying policies across state lines may also lead to inefficiencies and market imbalances. It also possible, of course, that some commitments made during periods when the economy was growing will be reconsidered as the economy struggles to recover from a recession.

CHAPTER CONCLUSION

The collective effect of local, state, and private sector efforts to respond to climate change is significant and should be encouraged, but such efforts are not likely to be sufficient or sustainable over the long term without a strong framework of federal policies and programs that ensure all U.S. stakeholders are working toward coherent national goals.

As described briefly in this chapter and explored in the ACC panel reports, there are already many efforts under way across the United States (led by state and local governments, and private sector and nongovernmental organizations) to reduce domestic GHG emissions, to adapt to anticipated impacts of climate change, and to advance systems for collecting and sharing climate-related information. Although there can be real benefits to having these actions take place in such a decentralized fashion,[20] in the judgment of the committee the many risks posed by climate change—coupled with the scale and scope of responses needed to respond effectively—demand national-level leadership and coordination. The appropriate balance between federal and nonfederal responsibilities depends on the domain of action. The ACC panel reports provide detailed discussion about the different types of federal leadership and coordination efforts that are most needed in the domains of advancing scientific understanding, limiting the magnitude of climate change, adapting to its impacts, and informing effective decisions.

CHAPTER TWO

Causes and Consequences of Climate Change

Climate change poses significant risks for a wide range of human and natural systems.

This statement, based on the conclusions of the America's Climate Choices (ACC) panel report *Advancing the Science of Climate Change*, stems from a substantial array of evidence and is consistent with the conclusions drawn in other recent scientific assessments, including reports by the U.S. Global Change Research Program (USGCRP) and the Intergovernmental Panel on Climate Change (IPCC). Although the scientific process is always open to new ideas and results, the fundamental causes and consequences of climate change have been established by many years of scientific research,[1] are supported by many different lines of evidence, and have stood firm in the face of careful examination, repeated testing, and the rigorous evaluation of alternative theories and explanations. This chapter provides a brief overview of some basic facts about the risks posed by climate change (see Box 2.1); additional explanation and detail can be found in the reports noted above.

OBSERVED CLIMATE CHANGE

Earth is warming.[2] The average temperature of the Earth's surface increased by about 1.4°F (0.8°C) over the past 100 years, with about 1.0°F (0.6°C) of this warming occurring over just the past three decades (see Figure 2.1). Warming has also been observed specifically in the lower atmosphere[3] and the upper oceans.[4] Additional, indirect indications of warming include widespread reductions in glaciers and Arctic sea ice,[5] rising sea levels,[6] and changes in plant and animal species.[7]

The preponderance of the scientific evidence points to human activities—especially the release of CO_2 and other heat-trapping greenhouse gases (GHGs) into the atmosphere —as the most likely cause for most of the global warming that has occurred over the last 50 years or so.[8] This finding is supported by numerous lines of evidence, including:

BOX 2.1
Climate Change and Risk

As used here, the term *risk* applies to undesired events that may occur in the future but are not certain to occur. Analysts typically quantify risks along two dimensions—the probability that an event will occur, and the magnitude or consequence(s) of the event—and multiply the two to get a risk estimate (probability times consequence).[a] Consequences, however, have many dimensions. They vary in terms of which human values and concerns they affect (lives, livelihoods, community integrity, nonhuman species, etc.), where they are likely to occur, whom they may affect, and when they are likely to cause harm. Consequences also vary in terms of perception and significance to those who face the risks—for example, in the degree to which the consequence is understood or evokes dread (unknown risks can sometimes concern people more than other risks), and even in the trust in the organizations that manage such risks (mistrust in the managing organizations tends to increase perceived risk).[b]

The risks posed by climate change are thus complex. As this chapter discusses, climate change drives a variety of biophysical processes, which leads to a variety of potential consequences for many things that people value. Risks will change over time, and consequences will be highly variable across different locations and population groups. Scientific analyses can improve understanding of the risks associated with climate change, including how different human reactions might change those risks and at what cost. One way this is done, for instance, is through scenario analyses that illustrate a range of possible future conditions and that can be used to test out the performance of different response strategies. But regardless of how much supporting scientific information is available, making choices about how to act in the face of uncertainty can prove contentious if people disagree about the nature of the risks they face or about which elements of these risks are most important.

---

[a] C. Jaeger, O. Renn, E. A. Rosa, and T. Webler, *Risk, Uncertainly and Rational Action* (London: Earthscan, 2001).

[b] P. Slovic, B. Fischhoff, and S. Lichtenstein, "Facts and fears: Understanding perceived risk," in *Societal Risk Assessment: How Safe Is Safe Enough?* (R. C. Schwing and W. A. Albers, Jr., eds. New York: Plenum, 1980); O. Renn, *Risk Governance: Coping with Uncertainty in a Complex World* (London: Earthscan, 2008).

- The concentration of CO_2 in the atmosphere has increased markedly over the past 150 years (see Figure 2.2) and is now higher than at any time in at least 800,000 years.[9]
- The long-term rise in CO_2 concentrations can be attributed primarily to the growth in human CO_2 emissions from fossil fuel burning (Figure 2.2), with deforestation and other land use and land cover changes also contributing.[10]
- Concentrations of other GHGs, including methane, nitrous oxide, and certain halogenated gases, have also increased as a result of human activities.[11]

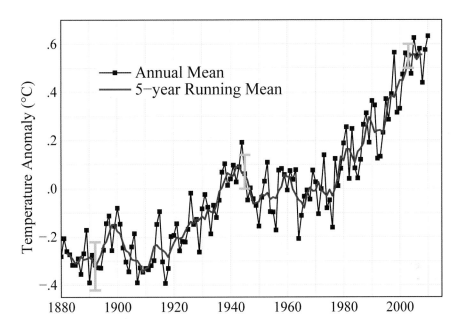

FIGURE 2.1 Global surface temperature change from 1880 to 2010, reported as a deviation from the 1951-1980 average. The black curve shows the globally and annually averaged near-surface temperature derived from a variety of instruments including thermometers, satellites, and various ocean sensors, all carefully calibrated and quality-controlled to remove errors. Green bars indicate the 95 percent confidence interval. The red curve shows a five year running average. The data show considerable year-to-year and decade-to-decade variability, but the long-term trend is clearly one of warming. SOURCE: NASA/GISS (other research groups find similar results; see, for example, *http://www.cru.uea.ac.uk/cru/data/temperature/*).

- Both basic physical principles and sophisticated models of the Earth's climate system definitively show that when the GHG concentrations increase, warming will occur.
- Careful analyses of observations and model results indicate that natural factors such as internal climate variability or changes in incoming energy from the sun cannot explain the long-term global warming trend.[12] Natural climate variability does, however, lead to substantial year-to-year and decade-to-decade fluctuations in temperature and other climate variables (as evident in Figure 2.1).

Global warming has been accompanied by a number of other global and regional environmental changes, which are broadly consistent with the changes expected in a warming world. However, establishing a direct, empirically-based causal link between

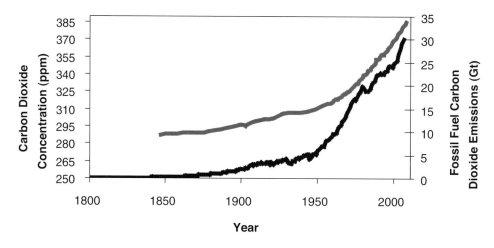

FIGURE 2.2 The concentration of CO_2 in the atmosphere (magenta line and left axis, in parts per million), as measured in ice cores and canisters of air collected from multiple locations around the globe, has risen steady since the mid-19th century, with the sharpest rate of increase occurring over the past few decades. Much of this increase can be attributed to global CO_2 emissions due to fossil fuel burning (blue line and right axis), which include estimated emissions from the production, distribution, and consumption of fossil fuels, plus a small contribution from cement production. Changes in land use and land cover—especially deforestation—also contribute to changes in atmospheric CO_2 concentrations, with current emissions estimated at 4.4 million metric tons per year (or about 12 percent of total emissions from human sources). SOURCE: NRC, *America's Climate Choices: Advancing the Science of Climate Change* (Washington, D.C.: National Academies Press, 2010). See Figure 2.3 from that report for further references.

global warming and these other changes is difficult: often the regional changes remain within the range of past observed variability, the data are not extensive enough, or the models not sufficiently developed to clearly identify an anthropogenic signal. As a result, only a few changes have been directly linked to human activities using formal scientific attribution methods.[13]

Among the ongoing changes in the physical climate system[14] that can be linked, at least in part, to increasing temperatures at the Earth's surface are widespread melting of glaciers and ice sheets,[15] rising global average sea levels,[16] and decreases in Northern Hemisphere snow cover[17] and Arctic sea ice.[18] These changes have, in turn, been linked to a number of impacts on other physical and biological systems over the past several decades.[19] For example, permafrost (permanently frozen ground) is thawing across many regions in the Northern Hemisphere,[20] lakes and rivers are freezing later and melting earlier.[21] Elevated CO_2 levels in the atmosphere are also causing widespread acidification of the world's oceans, which poses significant risks to ocean ecosystems.[22]

Changes in climate and related factors have been observed in the United States. These were recently assessed in *Global Climate Change Impacts in the United States*[23] and discussed in two of the ACC panel reports (NRC, *Advancing the Science* and *Adapting to the Impacts*), and include the following:

- U.S. average air temperature increased by more than 2°F over the past 50 years, and total precipitation increased on average by about 5 percent;[24]
- Sea level has risen along most of the U.S. coast, and sea level rise is already eroding shorelines, drowning wetlands, and threatening the built environment;[25]
- Permafrost temperatures have increased throughout Alaska since the late 1970s, damaging roads, runways, water and sewer systems, and other infrastructure;[26]
- There have been widespread temperature-related reductions in snowpack in the northeastern and western United States over the last 50 years, leading to changes in the seasonal timing of river runoff;[27]
- Precipitation patterns have changed: heavy downpours have become more frequent and more intense;[28] the frequency of drought has increased over the past 50 years in the southeastern and western United States, while the Midwest and Great Plains have seen a reduction in drought frequency;[29] and
- The frequency of large wildfires and the length of the fire season have increased substantially in both the western United States and Alaska.[30]

FUTURE CLIMATE CHANGE

Projections of future climate change impacts are developed in three steps:

(i) Emission Scenarios: Scientists first develop different scenarios of how GHG emissions and other human drivers of climate change (such as land use change) could plausibly evolve over the 21st century. Each scenario is based on specific assumptions about future social, economic, technological, and environmental change.[31]

(ii) Climate Simulation: Computer-based models of the climate system[32] are then used to estimate how temperature, precipitation, storm patterns, and other aspects of climate would respond to each emission scenario. Typically, a number of different scenarios and models are used to explore a wide range of possible future climate changes.

(iii) Impact Assessment: Finally, researchers evaluate the potential impacts of climate change, including their likelihood and temporal evolution by combining climate

model results with knowledge about the vulnerability[33] and adaptive capacity of various human and natural systems.

There are two major sources of uncertainty in future climate projections. One comes from the scenarios of emissions and other socioeconomic changes. Future emissions (and future vulnerability to climate change) will be determined by a complex set of developments taking place around the world—related to population, economic growth, energy, land use, technology and innovation, and other factors. It is not possible to predict how all such factors will change in the coming decades, but scenarios allow us to explore the implications of different pathways.

The second source of uncertainty is the response of the climate system to the increased concentration of GHGs, or "climate sensitivity." Even if future emissions were known exactly—that is, if a given emission scenario held true exactly—the magnitude of future climate change and the severity of its impacts cannot be predicted with absolute certainty, due to incomplete knowledge of how the climate system will respond. What is known with a high degree of certainty however, is the direction of the climate system's response to changes in GHG emissions: that is, reducing GHG emissions will lead to less warming and less severe impacts than if emissions are not reduced.

Significant changes are in store. The IPCC's assessment of future climate change projects that Earth's average surface temperature will increase (in the absence of new emissions mitigation policies) between 2.0 and 11.5°F (1.1 to 6.4°C) by the end of the 21st century, relative to the average global surface temperature during 1980-1999.[34] As discussed above, this range reflects the potential trajectories of future GHG emission rates as well as uncertainties in our understanding of the climate system response. A subset of these results is shown in Figures 2.3 and 2.4.

One notable feature of future climate projections is that the impacts of the differences among GHG emission scenarios grows with time. For example, the lowest and highest emission scenarios in Figure 2.3 lead to similar temperature changes over the next few decades, but very large differences in temperature by the end of the century.[35] This represents both a challenge and an opportunity—a challenge because emissions reductions that people make today will have little immediate effect on the climate; an opportunity because emissions reduction efforts made in the near term will affect climate outcomes many decades from now (see Box 2.2).

Two other notable features of the climate projections shown in Figures 2.3 and 2.4 relate to the temporal scales involved. First, the effects of GHG emissions can take

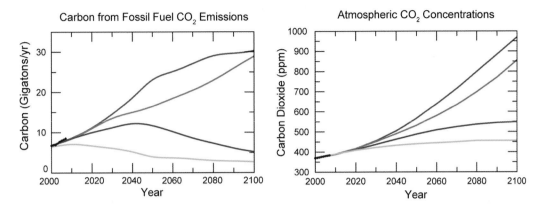

FIGURE 2.3 Observed (black curve) and projected (colored curves) changes in global CO_2 emissions (left, in gigatons of carbon) and atmospheric CO_2 concentrations (right) for four illustrative scenarios of future emissions. SOURCE: USGCRP, *Global Climate Change Impacts in the United States*, eds. T. R. Karl, J. M. Melillo, and T. C. Peterson (Cambridge, UK: Cambridge University Press, 2009) and model projections from CMIP3-A (G. A. Meehl et al., "The WCRP CMIP3 multimodel dataset: A new era in climate change research" [*Bulletin of the American Meteorological Society*, 88, 1383-1394, 2007]). The three scenarios illustrated here are based on IPCC/SRES (N. Nakicenovic et al., *Special Report on Emissions Scenarios, International Panel on Climate Change* [Cambridge, UK: Cambridge University Press, 2000]) scenarios: B1 (blue line, "lower emissions"), A2 (orange line, "higher emissions"), A1F1 (pink line, "even higher emissions"), and plus constant 20th century forcing (green line).

decades to fully manifest themselves. For example, in the "blue" scenario, emissions peak in 2040 but temperatures continue to increase through the end of the century. Second, climate changes caused by CO_2 persist for very long time scales. Figure 2.4 shows that the temperature perturbations produced by emissions in the 20th and early 21st century continue to warm the climate in 2100. In fact, the warming extends well beyond 2100—for CO_2, the time scale for such perturbations is millennial (e.g., some of the CO_2 we emit today is expected to remain in the atmosphere in the year 3000).[36]

Future climate change poses numerous known and unknown risks. The impacts of climate change—on coasts, water resources, agriculture, ecosystems, transportation systems, and other human and natural systems—can generally be expected to intensify with warming.[37] Some of these impacts are well understood and can be quantified with reasonable scientific confidence, while others are much less understood and can only be qualitatively described. A few examples of impacts that have been projected to occur across a range of future warming scenarios include:

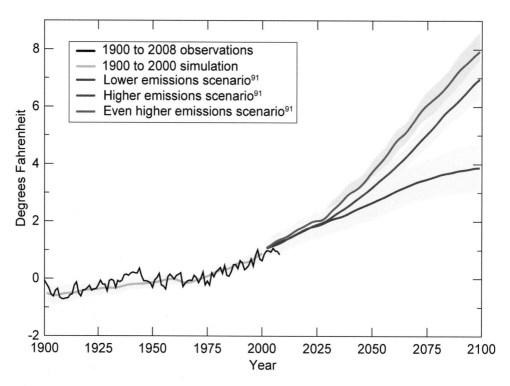

FIGURE 2.4 Observed (black curve) and projected (colored curves) changes in global average surface temperature for three of the illustrative scenarios of future emissions in Figure 2.3 (plus, in green, modeled 20th century climate). The shading around each curve indicates the range of central values produced by 15 different models using the same emission scenario (exact models used are listed in footnote #93 in USGCRP, *Global Climate Change Impacts in the United States*, eds. T. R. Karl, J. M. Melillo, and T. C. Peterson [Cambridge, UK: Cambridge University Press, 2009]). Other emissions scenarios and models show a substantially broader range of possible future temperature trajectories (see IPCC *Climate Change 2007: The Physical Science Basis. Contribution of Working Group I to the Fourth Assessment Report of the Intergovernmental Panel on Climate Change*, eds. S. Solomon, D. Qin, M. Manning, Z. Chen, M. Marquis, K. B. Averyt, M.Tignor, and H. L. Miller (Cambridge, UK: Cambridge University Press, 2007). SOURCE: USGCRP, *Global Climate Change Impacts*, based on observational data from Smith et al., "Improvements to NOAA's historical merged land–ocean surface temperature analysis (1880–2006)" (*Journal of Climate*, 21[10]: 2283-2296, 2008) and model projections from CMIP3-A (G. A. Meehl et al., "The WCRP CMIP3 multimodel dataset: A new era in climate change research" [*Bulletin of the American Meteorological Society*, 88: 1383–1394, 2007]), using the same IPCC emission scenarios as in Figure 2.3.

- more intense, more frequent, and longer-lasting heat waves, both globally[38] and in the United States (see Figure 2.5);
- global sea level rise[39] with potentially large effects on infrastructure, beach erosion, loss of wetlands, vulnerability to storm surge flooding in the Gulf Coast and other coastal regions,[40] and irreversible commitments to future

changes in the geography of the Earth as many coastal and island features ultimately become submerged;[41]

- widespread bleaching and stresses on coral reefs, globally[42] and in the Florida Keys, Hawaii, and U.S. island possessions,[43] due to the combined effects of heat stress, ocean acidification,[44] pollution, and overfishing;
- greater drying of the arid Southwest (putting additional pressure on water resources) and expansion of deserts in the United States;[45]
- effects on agriculture due to elevated CO_2 levels, temperature and precipitation changes, and also by possible increases in weeds, diseases, and insect pests;[46]
- shifts in the ranges of forest tree-species (northward and upslope), increases in forest fire risk across much of the western United States,[47] and a potential increase in the number of species at risk of extinction;[48] and
- increased potential of public health risks, for instance, from heat stress; from elevated ozone air pollution; from certain diseases transmitted by food, water, and insects; and from direct injury and death due to extreme weather events.[49]

Climate change will affect specific regions and segments of society differently because of varying exposures and adaptive capacities. For instance, public health threats and outcomes are affected not only by climate factors, but also by factors such as wealth and lifestyle, status of public health systems, and access to medical care and information. As another example, coastal cities that have instituted measures to protect critical infrastructure (for utilities, transportation, etc.) may be less vulnerable to the impacts of sea level rise and storm surges.

The physical and social impacts of climate change are expected to have substantial economic implications throughout the United States, but these effects will be unevenly distributed across regions, populations, and sectors.[50] Quantitatively estimating economic impacts is controversial, due to the uncertainties in climate change impact projections themselves, and to the lack of sound methodologies for assigning economic value to many key impacts, especially nonmarket costs such as loss of ecosystem services and spillover costs occurring as a result of climate change impacts elsewhere in the world.

In addition to the potential impacts that we are able to identify today, there is a real possibility of impacts that have not been anticipated. This possibility, coupled with our limited ability to predict the timing and location of some climate-related impacts, and our incomplete understanding of the vulnerabilities of different populations and sectors will make adaptation to climate change especially challenging.

BOX 2.2
Emissions Headroom

Although the question of what constitutes a "safe" level of climate change remains a matter of active debate, the United States and much of the international community have expressed support for the goal of limiting global temperature rise relative to the pre-industrial times to no more than 2°C (3.6°F). This target is often stated to be equivalent to limiting global atmospheric CO_2 concentrations to no more than 450 ppm. Because of uncertainties in climate sensitivity, however, this is not a precise relationship. The best estimate from climate models is that there is a 50 percent probability of limiting global temperature increase to 2°C or less if CO_2 concentrations are not allowed to rise above 450 ppm.[a]

Another, and perhaps more useful, way to view the problem is in terms of total cumulative CO_2 emissions (i.e., the sum of all emissions over time). How much greenhouse gas can be emitted and still keep the global temperature rise below 2°C? A rough estimate can be obtained using the near-linear relationship that exists between the cumulative carbon emissions from human activities since the Industrial Revolution and the long-term rise in the Earth's average surface temperature.[b] Based on this relationship, it is estimated that keeping global temperature rise within 2°C requires limiting cumulative emissions to approximately 4000 billion tons of CO_2. There is considerable uncertainty in this result however, with the cumulative emissions likely ranging from about 2900 to 5800 billion tons. Cumulative human-related emissions since the Industrial Revolution total about 1800 billion tons of CO_2. This leaves an "emissions headroom" (i.e., the remaining amount that can be emitted) of somewhere between 1,100 to 4,000 billion tons of CO_2, with a central estimated headroom of roughly 2,200 billion tons.

The world currently emits ~30 billion tons of CO_2 per year from fossil fuels, with one-fifth of this amount emitted by the United States. If global emissions continued at that rate, the central estimated headroom would be used up (as a rough approximation) somewhere in the range of 40 to 130 years, with a most probable value of roughly 70 years—after which point emissions would have to drop to zero. Thus the degree of headroom is uncertain—there could be very little left, or there could be a significant amount, depending mainly on uncertainties in the climate sensitivity.

These estimates are undoubtedly too optimistic however, because without policy intervention,

CHAPTER CONCLUSION

Although the exact details cannot be predicted with certainty, there is a clear scientific understanding that climate change poses serious risks to human society and many of the physical and ecological systems upon which society depends—with the specific impacts of concern, and the relative likelihood of those impacts, varying significantly from place to place and over time. It is likewise clear that actions to reduce GHG emissions and to increase adaptive capacity will lower the likelihood and the consequences of these risks.

global emissions will not continue at current rates, but rather, will continue to rise. For instance, NRC, *Climate Stabilization Targets* projects that total global emissions between 2009 and 2050 will exceed 1,000 billion tons of CO_2 under a scenario with no new policy interventions. If the headroom is at the lower end of the range listed above, it would all be used up by 2050 (and, of course, emissions are highly unlikely to drop to zero after that point). In addition, the available headroom shrinks if the goal for limiting global average temperature rise is more stringent than the 2°C target. These uncertainties illustrate how America's climate choices fundamentally involve judgments and perceptions about acceptable risk.

Commitments to emissions in the form of the infrastructure investments will be key in determining how quickly we use up the emissions headroom. The world has already locked in a large amount of future emissions through both old and recent investments in capital stock (e.g., cars and trucks, home furnaces and boilers, building shells, chemical plants and factories, power plants) and fixed infrastructure investments. One study estimates that if the world were to build no further energy-using stock, while letting every existing fossil fuel-using device reach the end of its useful life without modification, somewhere in the range of 280-700 billion tons of CO_2 would be emitted.[c] How much is emitted above that amount depends critically upon the types of energy infrastructure the world invests in during the coming decades. If much of the new capital stock is powered by fossil fuels, then the emissions headroom will be rapidly depleted. If, on the other hand, concerns over climate change lead to concerted actions to slow the addition of new fossil-fuel-using capital stock, or if existing stock were retired early or retrofit (e.g., retrofits of coal plants to capture and store CO_2), the headroom will last longer.

---

[a] IPCC, *Climate Change 2007: The Physical Science Basis. Contribution of Working Group I to the Fourth Assessment Report of the Intergovernmental Panel on Climate Change*, eds. S. Solomon, D. Qin, M. Manning, Z. Chen, M. Marquis, K. B. Averyt, M.Tignor, and H. L. Miller (Cambridge, UK: Cambridge University Press, 2007)

[b] NRC, *Climate Stabilization Targets: Emissions, Concentrations, and Impacts over Decades to Millennia* (Washington, D.C.: National Academies Press, 2010); and H. D. Matthews, N. P. Gillett, P. A. Scott, and K. Zickfeld, "The proportionality of global warming to cumulate carbon emissions" (*Nature* 459[7248]:829-U823, 2009). Note that in *Stabilization Targets*, numbers are given in tons of carbon, whereas here they are converted to tons of CO_2.

[c] S. J. Davis, K. Calderia, and D. Matthews, "Future CO_2 emissions and climate change from existing energy infrastructure" (*Science* 10 329[5997]:1330-1333, 2010, doi: 10.1126/science.1188566).

Waiting for unacceptable impacts to occur before taking action is imprudent because many of the impacts of GHGs emitted today will not fully manifest themselves for decades; and once they do appear, they can be with us for hundreds or even thousands of years. The amount of warming is expected to increase with the cumulative amount of GHGs emitted, and thus the chances of encountering dangerous climate impacts grows with every extra ton we emit. At the same time, national and world demand for energy is on the rise, and new investments in energy infrastructure are inevitable. If those investments are in CO_2-emitting infrastructure, we will have committed ourselves to growing GHG emissions for decades to come.

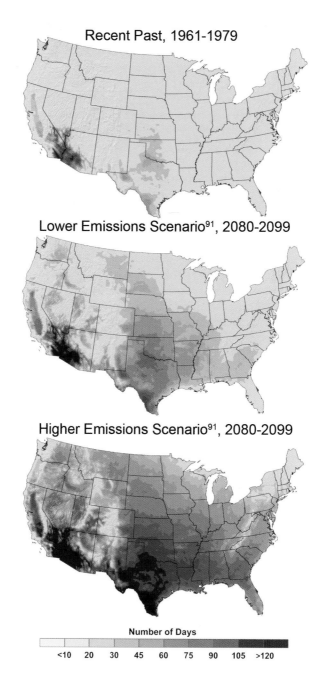

Recent Past, 1961-1979

Lower Emissions Scenario[91], 2080-2099

Higher Emissions Scenario[91], 2080-2099

Number of Days

<10 20 30 45 60 75 90 105 >120

FIGURE 2.5 The number of days per year in which temperatures are projected to exceed 100°F by late this century compared to the 1960s and 1970s under two different scenarios of future GHG emissions (IPCC SRES scenarios B1 and AIF1, illustrated in Figure 2.3). SOURCE: USGCRP *Global Climate Change Impacts*, p. 90.

Thus in the judgment of the committee, the environmental, economic, and humanitarian risks of climate change indicate a pressing need for substantial actions to limit the magnitude of climate change and to prepare for adapting to its impacts. Undertaking such actions will require making choices in the face of incomplete and imperfect information about the future.

CHAPTER THREE

The Unique Challenges of Climate Change

The climate system is highly complex, as are the human institutions that are affected by and that must respond to climate change.

The difficulty of developing sound strategies for responding to climate change, and of building public support for such strategies, stems in part from the inherent complexity of the issue. Some of this complexity relates to the physical science of climate change; but understanding and responding to climate change also raises many social, economic, ethical, and political challenges. The chapter highlights some of the unique challenges posed by climate change that must be considered in designing the nation's response strategies.

There are complex linkages among emissions, concentrations, climate changes, and impacts. Projecting future climate change requires understanding numerous linkages among human activities, greenhouse gas (GHG) emissions, changes in atmospheric composition, the response of the climate system, and impacts on human and natural systems. The basic links in this chain are well understood, but some elements (in particular, projecting specific impacts at specific times and places) are much less so. As a result, the outcomes of actions to reduce emissions or to reduce the vulnerabilities of human and natural systems must often be presented in probabilistic or qualitative terms, rather than as certain predictions.

Lack of certainty about the details of future climate change is not, however, a justification for inaction. People routinely take actions despite imperfect or incomplete knowledge about the future in situations such as buying home insurance, saving for retirement, or planning business strategies. Likewise, people use probability data from weather forecasts to decide if they should take an umbrella to work, move a scheduled outdoor event indoors, or cancel a ball game. Indeed, it could be argued that uncertainty about future climate risks is a compelling reason *for* taking proactive steps to reduce the likelihood of adverse consequences.

There are significant time lags in the climate system. It takes very long time periods (decades to millennia) for some aspects of the climate system to respond fully to changes in atmospheric GHG concentrations.[1] This is because the world's oceans can store a large amount of heat—so it takes a long time for the climate system to warm up in response to changes in GHG concentrations[2]—and because impacts such as sea level rise and the melting of ice sheets can take several centuries or even millennia to be fully expressed. Some GHGs (such as methane) are removed from the atmosphere within about a decade, but CO_2 persists much longer—approximately 20 percent of the CO_2 emitted today will remain in the atmosphere more than a millennium from now.[3] Thus, a failure to reduce GHG emissions in the near-term will "lock in" a certain amount of future climate change for decades, if not centuries, to come.

There are also significant time lags in human response systems. GHG emissions are to a large extent built into societal infrastructure (e.g., buildings, power plants, settlement and transportation patterns) and into human habits and organizational routines, few of which change quickly. Market incentives affecting capital investments leave little room for considering consequences on century or longer time scale. Nevertheless, making major reductions in GHG emissions and preparing to adapt to the effects of climate change will require transformative changes, for instance, in how the country produces and uses energy (see Box 3.1), builds buildings and transportation infrastructure, and manages water and other natural resources. It will likewise require significant changes in consumer choices, travel behavior, and other individual and household-level decisions. Overcoming the inertia of the status quo in advancing these sorts of transformations will pose challenges for government, industry, agriculture, and individual citizens alike.

An issue of particular concern is that much of the equipment and infrastructure that leads to GHG emissions (e.g., roads, vehicles, buildings, power plants) have lifetimes of decades. There are often strong economic pressures to continue use of such equipment and infrastructure, rather than retrofitting or replacing with a lower-emitting option. Making substantial emission reductions within the next few decades will require accelerating this turnover faster than projected business-as-usual rates.[4]

Risks, judgments about risk, and adaptation needs are highly variable across different contexts. Different regions, economic and resource sectors, and populations will experience different impacts from climate change, will vary in their ability to tolerate and adapt to such impacts, and will hence differ in their judgments about the potential risks posed by climate change. For instance, coastal communities that are vulnerable to serious disruptions could be expected to view the risks of climate change as quite serious. Actions that are taken in response to climate change will also pose differing types of risks to different regions, sectors, and populations. For instance,

---

BOX 3.1
The U.S. Energy System

The U.S. energy system includes a vast and complex set of interlocking technologies for the production, distribution, and use of fuels and electricity.[a] This includes technologies that convert primary energy resources (e.g., nuclear energy, renewable sources such as solar and wind, and the fossil fuels coal, oil, and natural gas) into useful forms such as gasoline and electricity; technologies that transmit this energy to consumers (e.g., electrical transmission and distribution systems, gas pipelines); technologies that store or utilize this energy (e.g., batteries, motors, lights, home appliances); and associated demand-side technologies that control energy use (e.g., advanced electricity metering systems). Another key component of this system is the people that use the energy—their behaviors and preferences play a major role in shaping energy technologies.

Currently, the United States relies on carbon-based fossil fuels for more than 85 percent of its energy needs. This dependence evolved not only because fossil fuels were available at low market costs but also because their physical and chemical properties are well suited to particular uses: petroleum for transportation; natural gas as an industrial feedstock, for residential and commercial space heating, and more recently as a fuel for electric power generation; and coal for the generation of electricity and as a feedstock for some industrial processes. Indeed, almost all consumer-based, industrial, and governmental activities require the consumption of fossil fuels, either directly or indirectly.

Absent strong and sustained policy intervention, fossil fuels are projected to remain the nation's primary source of energy for the foreseeable future. Compared with alternative sources of energy, fossil fuels would likely remain relatively inexpensive to produce, and they would continue to benefit from past investments in vast existing infrastructure—investments that would need to be duplicated (in whole or in part) to enable wide-scale displacement by alternative energy sources. The nation's reliance on carbon-based fossil fuels would only be significantly reduced in the near-term if the prices of those fuels were increased to reflect the full social costs of their extraction, transformation, distribution, and use; and only if there are incentives to encourage research and development aimed at reducing the cost and promoting the commercialization of alternative energy sources.

---

[a] The material in this box was adapted from NRC, *America's Energy Future: Technology and Transformation: Summary Edition* (Washington, D.C.: National Academies Press, 2009).

---

individuals and organizations that are heavily invested in carbon-intensive industries may prefer to face the risks of climate change impacts rather than face the potential costs of policies to limit GHG emissions. Decision makers will thus inevitably face some difficult choices and trade-offs in seeking to protect the interests of different constituencies.

Decisions affecting climate change are made at all levels of society. The federal
government can play a critical leadership role in setting policies that affect the actions
of all parts of society. But much of the responsibility and opportunity for responding
to climate change rests with state and local governments and with the private sector (which accounts for most of the nation's capital investments, industrial production, and employment). Decisions made at the individual and household level also
play a major role in driving GHG emissions, and of course, public support is critical
for motivating political leaders to take actions in response to climate change. A U.S.
strategy for responding to climate change must therefore include careful consideration of which information, incentives, and regulations (provided by which level of
government) will most effectively engage and facilitate wise decision making by these
multiple actors. In some cases, the appropriate federal role may be limited to decision
support, while in other contexts, more active policy guidance and coordination power
are needed.

Limiting climate change requires global-scale efforts. A molecule of CO_2 emitted in
India or China has the same effect on the climate system as a molecule emitted in the
United States. There is wide agreement that limiting the magnitude of climate change
will require substantial action on the part of all major GHG-emitting nations, including
both the industrialized nations and the rapidly developing countries whose relative
share of global emissions is rapidly increasing (see Figure 3.1). Yet there are many different perspectives on how to define each country's responsibilities for contributing
to the global effort.[5] Some argue that U.S. action must be conditioned on actions by
other nations, given the economic disadvantages that the country might face if it committed to significant emission reductions without similar commitments from other nations. Others argue that the United States, as the country with largest historical share
of GHG emissions and with one of the highest per capita GHG emission rates, has an
ethical obligation to substantially reduce domestic emissions, even in the absence of
commitments from other nations. Still others suggest that there will be substantial
economic advantages in leading the development of new technologies to deal with
climate change. There is no simple way to reconcile these different views, but it is clear
that strong, credible U.S. policies for reducing domestic emissions will help advance
international-level efforts to do the same.

Climate change is one of multiple, interconnected challenges. Climate change is just
one of many interacting factors affecting humans and their environment. Coastal environments, for example, are being affected not only by GHG-driven changes such as sea
level rise, ocean acidification, changes in air and water temperature, and precipitation
and storm patterns, but also by pollution runoff, invasive species, coastal development,

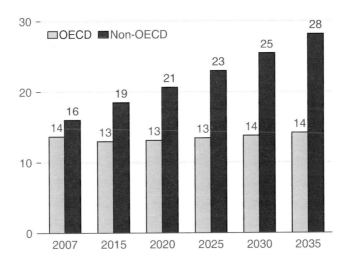

FIGURE 3.1 World energy-related CO_2 emission projections (in billion metric tons CO_2), by OECD (Organization for Economic Cooperation and Development) and non-OECD countries over the period 2007-2035. Non-OECD countries include developing, newly industrialized, and Eastern European and former Soviet countries. For a list of OECD and non-OECD countries, see: *http://www.oecd.org/countrieslist/*. SOURCE: Energy Information Administration / International Energy Outlook. 2010.

and overfishing. These different issues are often studied and managed as isolated matters, without recognizing and accounting for interconnected causes and interactive effects. Likewise on a broader global scale, many different issues affect and are affected by climate change—such as food production, water supplies, human health, energy production and use, economic development, security concerns—but these are seldom addressed in an integrated manner.

These sorts of inter-linkages not only pose difficult challenges, but also offer important opportunities for alleviating multiple problems simultaneously. For instance, integrated management plans for protection of coastal zones can help alleviate many of the climate-related and non-climate-related concerns listed above. Actions taken to reduce fossil fuel use can offer substantial benefits for human health (by reducing emissions of conventional air pollutants) and for national security (by reducing dependence on imported energy sources).[6]

The costs and benefits of different courses of action are generally not well known.
Decision making often involves weighing the possible benefits and costs of one course of action against another. Decisions on actions to limit or adapt to climate

change have characteristics that make such analyses extremely challenging. For example:

- **Costs and benefits are difficult to quantify.** It is difficult to characterize the costs and benefits of climate change impacts in part because many of the natural assets and ecosystem services that could be affected by climate change have no market value or are priced in a way that does not truly reflect social values. The costs of efforts to limit or adapt to climate change are often perceived as being comparatively more certain, but in reality, different assumptions (about, for example, the rate at which new technologies are developed and brought to market) can result in widely varying conclusions about the costs of such actions.[7]

- **Costs of actions to limit climate change risks are immediate, but many benefits will occur elsewhere and affect future generations.** Although the United States is vulnerable to many impacts of climate change, other parts of the world—for example, low-lying island nations—are at a greater risk of catastrophic impacts[8] and thus would benefit most from actions to limit climate change. Even in this country, the primary beneficiaries of near-term actions to reduce climate change risks are the future generations that would avoid severe impacts later in the century and beyond (although there are some immediate, local benefits, as discussed below). In economic analyses, this issue is usually addressed by choosing a rate at which future benefits and costs are discounted relative to current benefits and costs. But many economists and others have expressed concerns about using these conventional discounting techniques to value public benefits, especially in the context of climate change, where trade-offs must be evaluated across multiple generations.[9]

- **Collateral costs and benefits also need to be considered.** The Earth's physical systems (atmosphere, oceans, land surface, fresh water), ecological systems, and human social systems are highly interconnected. Changes in one system can affect others, and actions taken to limit or adapt to climate change may result in unintended consequences, both positive and negative. For instance, reducing the use of fossil fuels to limit GHG emissions can offer the ancillary benefit of also reducing emissions of several health-damaging air pollutants (e.g., nitrogen oxides, sulfur dioxide, particulate matter, mercury).[10] Improving freshwater use efficiency to increase a community's resilience to climate change can also help that community deal with natural variations in water supply. As an example of ancillary costs—the production of biofuels for renewable energy has indirect effects on land use, which in turn can increase GHG emissions and negatively affect biodiversity.[11] Although some ancillary costs and benefits can be anticipated, there are inherent limits in the predictability of many such interactions, and it is thus difficult to account for them when assessing costs and benefits of specific climate change response actions.

Many factors complicate and impede public understanding of climate change.[12] Public understanding of climate change is important because public opinion underpins policy and because the public—as consumers, employers, and community members—can initiate, implement, and support actions to reduce GHG emissions and encourage adaptation. Fully understanding climate change is a difficult task even for scientific experts using voluminous data and complex mathematical models. People who have less experience with quantitative data and less time to develop such detailed understanding must rely on other sources that may or may not provide trustworthy information. For instance, personal experience powerfully influences people's understanding of their environment. But this can be misleading in the context of climate change because long-term change is difficult to detect against natural variability without sustained systematic measurement, and because judgments of varying phenomena are strongly influenced by memorable and recent extreme events.[13]

People use different types of "mental models" to understand complex phenomena, and some of the prevalent models used for understanding climate change are inconsistent with scientific knowledge. For instance, many people appear to conflate GHGs with other forms of air pollution, such as particulates of sulfur or nitrogen oxides that (unlike long-lived GHGs) dissipate quickly when emissions are reduced.[14] Even well educated nonspecialists, including many science undergraduates, tend to systematically underestimate the degree to which CO_2 emissions must be reduced to stabilize atmospheric concentrations.[15]

Most people rely on secondary sources for information, especially the mass media; and some of these sources are affected by concerted campaigns against policies to limit CO_2 emissions, which promote beliefs about climate change that are not well-supported by scientific evidence. U.S. media coverage sometimes presents aspects of climate change that are uncontroversial among the research community as being matters of serious scientific debate.[16] Such factors likely play a role in the increasing polarization of public beliefs about climate change, along lines of political ideology, that has been observed in the United States.[17]

There will inevitably be additional risks and surprises. The climate system and human institutions that are affected by climate are exceedingly complex; consequently, it is impossible to anticipate all changes that may occur. Among the potential surprises that could be in store is that climate change turns out to be more manageable than

---

BOX 3.2
The Motivation for Action

The committee's judgment—that the environmental, economic, and humanitarian risks posed by climate change indicate a pressing need for substantial action to limit the magnitude of climate change and to prepare for adapting to its impacts—rests on numerous lines of argument alluded to in this chapter, including the following:

- There are significant time lags in how the climate system will respond to forcing from GHGs in the atmosphere, and there are likewise significant time lags in many of the social, technological, and political systems that must respond to climate change. Waiting to act until all uncertainties are resolved, or until impacts of concern have become fully manifest, will likely mean it is too late to have any meaningful effect in mitigating many risks.

- Due to uncertainties in climate sensitivity and other factors, one cannot say exactly how severe climate change and its impacts will be for any given level of atmospheric GHG concentrations. But even "moderate" climate change can pose serious risks, and there is the possibility of irreversible tipping points in the earth system, beyond which some particularly adverse impacts can occur.

- The sooner that serious efforts to reduce GHG emissions proceed, the less pressure there will be to make steeper (thus likely more expensive) emission reductions later on.[a]

- Both private and public sector decision makers face investment choices today that will affect the ability to limit emissions and to adapt to climate change for many years to come. For instance, investments in infrastructure for energy production and use can entail a massive commitment to future GHG emissions. Getting the relevant strategies and policies in place as soon as possible will provide crucial guidance for these investment decisions.

- Finally, in the committee's judgment, the risks associated with doing business as usual are a much greater concern than the risks associated with engaging in ambitious but measured response efforts. This is because many aspects of an "overly ambitious" policy response could be reversed or otherwise addressed, if needed, through subsequent policy change, whereas adverse changes in the climate system are much more difficult (indeed, on the time scale of our lifetimes, may be impossible) to "undo."

---

[a] L. Clarke, J. Edmonds, V. Krey, R. Richels, S. Rose, and M. Tavoni, "International climate policy architectures: Overview of the EMF 22 International Scenarios" (*Energy Economics* 31 (Supplement 2):S64-S81, 2009).

current projections suggest. (For instance, if climate sensitivity is on the low end of current estimates, then the resulting impacts might feasibly be adapted to without major costs or disruptions.) It is also quite possible, however, that climate change could be much more severe, or have much more severe impacts, than the average values in current projections. For example, a climate change-induced crop failure or severe drought could precipitate a geopolitical crisis.[18] There may also be "tipping points" in the climate system or affected human or natural systems, whereby a small incremental change pushes the system into a sudden and radical shift.[19] Currently, it is impossible to predict where or when such crises, tipping points, or other surprises might occur. It is worth noting, however, that the potential impacts associated with larger magnitudes of climate change are less well studied than more moderate climate change; and thus the potential for surprises is comparatively greater with larger magnitudes of warming.

CHAPTER CONCLUSION

The many complex characteristics of climate change discussed here—which reach across scientific, political, economic, psychological, and other dimensions—are not problems that must be fully "solved" before one can move ahead with making choices and taking action to address climate change (see Box 3.2). Rather, these are inherent features of climate change that must be recognized and understood in order to craft sound response strategies. As discussed later in this report, many possible response actions could be viewed as common-sense investments in our nation's future regardless of the complexities and uncertainties involved.

The issues highlighted in this chapter point to the idea that conventional analysis tools that have historically been used for guiding responses to major environmental problems are not well suited for addressing the complexities of climate change.[20] Instead, there is a need for decision frameworks that allow decision makers to weigh trade-offs, to act in the face of incomplete information, and to learn and adjust course over time. In the following chapter, we discuss the type of framework that is best suited for this context.

CHAPTER FOUR

A Framework for Making America's Climate Choices

Iterative risk management is a flexible and powerful approach for addressing the complex challenges of climate change.

Chapter 2 reviewed what is known about the risks posed by climate change, concluding that there is strong motivation for moving ahead with proactive response efforts. Yet as discussed in Chapter 3, the many complexities inherent to climate change make it difficult to define the specific actions that are needed in an effective long-term response strategy. This chapter introduces iterative risk management as an approach that lets decision makers begin to address climate change now, in a systematic way, while allowing response strategies to be adjusted and improved as new information and knowledge are gained. Iterative risk management is, in principle, a fairly simple and straightforward concept (see example in Box 4.1); however, the details of how it is actually applied in various real-world situations depend strongly on the context of that situation, including the specific problem being addressed, the stakeholders involved, the values and priorities of those stakeholders, and the decision-support tools and resources available. Thus, in this chapter we explore how iterative risk management may be used to address climate change in a general sense, but we do not attempt to offer a detailed formula for how to apply this framework in specific situations.

AN ITERATIVE RISK MANAGEMENT APPROACH TO CLIMATE CHANGE

As noted in Chapter 2, the risks posed by climate change are diverse and in almost all cases are imperfectly understood. Risk management involves deciding what to do in light of this imperfect information. Of course, one option is always to do nothing. Most everyone ignores some risks in daily life, and the United States might chose to give little attention to the risks associated with climate change. In the committee's view, however, such a path would not be prudent. Uncertainty is, after all, usually a two-edged sword: it is possible that future climate-related risks will be less serious than currently thought, but it is also possible that they will be even more serious. Even the most aggressive possible response could not remove all potential risks, since the

---

BOX 4.1
A Problem of Risk Management

The problem of managing climate-related risks shares important features with the problem faced by the captain of an ocean liner who had to pass through an iceberg-filled section of the ocean at night in the days before radar. The captain may have information about the location of some icebergs, but not all, and new ones can form at any time. The maneuverability and hull strength of the ocean liner—that is, its ability to avoid or survive a collision with an iceberg—may likewise be known in theory, but not tested in practice. Thus the risks are significant, but information is limited.

The captain could choose to go full-steam-ahead and hope that information becomes available in time to detect and avoid risks. Or the captain could consider alterative actions, such as taking a longer course through iceberg-free waters or fortifying the ship's hull—but there may be substantial costs associated with such actions. In any of these cases, it would be prudent to post lookouts to learn as much as possible about the risks ahead, to constantly evaluate the ship's environment and performance, and to be prepared to change course if needed, knowing that evasive maneuvers take time. In addition, it is essential to prepare for adverse outcomes that may occur, despite efforts to reduce their likelihood. The captain, in short, faces a problem of risk management.

America's climate choices are not, of course, made by one "captain," but by decision makers at all levels of society—from the President and Congress, to state and local leaders, to individual households and business owners. Nevertheless, the collective ship of state is best guided by coherent national strategies for assessing options and taking advantage of opportunities to reduce risk.

---

world is already committed to some degree of climate change as a result of GHG emissions to date.

Making America's climate choices thus necessarily involves managing risks that may be quite substantial and that cannot be eliminated, yet are often difficult to assess precisely. Making choices under such conditions can seem very difficult in the abstract, yet most people make such decisions every day. For instance, people decide how fast to drive, knowing that driving faster saves time but also uses more gas, increases the chances of a speeding ticket, makes an accident more likely, and makes the consequences of an accident more severe. People invest in measures to prevent fires in their homes and businesses, and they take out insurance to deal with the consequences in case fires do occur. People who make financial investments usually diversify their portfolios to hedge against uncertain future market changes. At the national level, history contains countless examples of policy makers taking action to address serious but poorly defined risks that could be neither eliminated nor responsibly ignored. For instance, investments in deterrence during the Cold War were justified as reducing the risk of a nuclear war, and investments in civil defense were justified as reducing the risks of catastrophic outcomes in case a war did occur.

In the case of climate change, appropriate strategies for reducing risks will change over time in light of new information, and so too will investments in different types of action. The committee suggests that some essential elements of a sound risk management strategy for responding to climate change include:

- Enacting policies and programs that reduce risk by limiting the causes of climate change and reducing vulnerability to its impacts;
- Investing in research and development efforts that increase knowledge and improve the number and effectiveness of response options;
- Developing institutions and processes that ensure pertinent information is collected and that link scientific and technical analysis with public deliberation and decision making; and
- Periodically evaluating how response efforts are progressing, and updating response goals and strategies in light of new information and understanding.[1]

To some extent, it is possible to make substitutions or trade-offs among investments in different elements of climate change response. For instance, substantially limiting the magnitude of climate change could make it less important to invest in adaptation efforts (recognizing that the outcomes of these different types of actions can occur over widely differing temporal and spatial scales, thus complicating direct trade-off relationships). In general, however, because the long-term benefits of investing in any particular response (e.g., R&D investments, emissions mitigation efforts, adaptation planning) are uncertain, a strategy of diversification across different types of responses will reduce risk more and be more robust than pursuing a single approach at the expense of all others.

Decision Frameworks for Addressing Climate Change

Historically, humans have responded to changing environments by a process of *muddling through;* that is, by taking an ad hoc approach to decision making as choices arise.[2] In the modern era, techniques and approaches have been developed that allow decision makers to think through complex issues systematically. One prominent approach is the *precautionary principle,*[3] which emphasizes avoidance of potentially serious or irreversible environmental harm, even when scientific uncertainties may be substantial. At the other extreme is what might be called *"staying the course,"* or not taking any action until the need for action is fully established and the consequences of any action are fully understood. Another common approach is *cost-benefit analysis* and other related instruments that attempt to weigh the potential outcomes of taking (or not taking) action using a common metric, usually dollars discounted to present values.

All of these approaches present serious drawbacks in the context of climate change. In muddling through, for instance, decisions are generally driven by immediate events and the lessons learned from one's most recent experiences. Such an approach makes it difficult to thoughtfully consider long-term consequences of climate-related decisions. For instance, with regard to the prospect of irreversible or "tipping point" impacts, it will be too late to change course if one waits until such impacts have begun to unfold.

Analyses based on the precautionary principle or staying the course both reflect a substantial aversion to risk. In the case of the precautionary principle, the goal is to minimize risks of future adverse consequences of climate change with little regard for present costs. In the case of staying the course, the goal is to minimize the risks of incurring costs from responding to climate change with little regard for the risks of climate change. These approaches do not provide a way to decide among competing goals (e.g., minimizing risks of climate change impacts versus minimizing risks to economic growth) or to deal systematically with uncertainty.[4]

Cost-benefit analysis has been applied in many evaluations of climate change policy[5] and can provide some useful insights in some contexts. But using cost-benefit analyses as a primary basis for making climate choices is problematic for a number of reasons. Many of the costs of climate change impacts are difficult or impossible to quantify.[6] The sheer diversity and extent of potential costs and benefits of climate change make it very difficult to aggregate costs. Estimates can vary widely, depending on normative judgments about risk aversion and about how to account for equity concerns across generations, social groups, and regions of the world.[7] Estimating the costs of actions to address climate change, while seemingly a more tractable task, is also problematic—for instance, because the costs of emission reductions over the coming decades depend critically on the pace of technological change.[8]

An *iterative risk management approach*[9] for making climate change-related decisions overcomes many of these limitations. This approach can draw upon multiple forms of input—including analyses used under precautionary principle and cost-benefit frameworks—but it is not limited to single criterion (such as risk avoidance or economic efficiency) for making choices. Iterative risk management is a system for assessing risks, identifying options that are robust across a range of possible futures, and assessing and revising those choices as new information emerges. In cases where uncertainties are substantial or risks cannot be reliably quantified, one can pursue multiple, complementary actions—sometimes called a "portfolio approach" or "hedging strategy." And ideally, this approach includes mechanisms for integrating scientific and technical analysis with broad-based deliberations among the stakeholders most affected by any given decision (see Box 4.2 on analytic deliberation processes).

NRC, *Informing Effective Decisions* emphasizes some key features of an iterative risk management process:

- It is not a single set of judgments at some point in time, but rather a process of ongoing assessment, action, reassessment, and response—which in the case of many climate-related decisions may persist for decades or longer.
- Eliminating all potential risks is impossible, as even the best possible decision will entail some residual risk. Determining which risks are acceptable or unacceptable is an integral part of the process of risk management. Different stakeholders will inevitably hold different views.
- For addressing a problem as complex as climate change, risk management should be implemented through a process of "adaptive governance" that involves assuring adequate coordination among the institutions and actors involved in responding to climate change, sharing information with decision makers across different levels and sectors, ensuring that decisions are regularly reviewed and adjusted in light of new information, and designing policies that can adapt but that are also durable over time. These concepts are illustrated in Figure 4.1 and discussed further in Chapter 5.

Similar principles have been recommended and illustrated by other high-level advisory groups worldwide, including, for instance, the IPCC, the United Nations Development Programme, the World Bank, the Australian Greenhouse Office, and the UK Climate Impacts Programme.[10] Closer to home, a number of NRC and other reports have pointed to the planning efforts being carried out by New York City as a good example of a climate change response strategy that embodies many key elements of iterative risk management.[11]

These efforts—which are set forth in PlaNYC, the city's sustainability and growth management initiative—include for instance:

- ambitious goals for limiting greenhouse gas (GHG) emissions, and a series of policies and programs to accomplish those goals (for instance, by reducing energy consumption by the city's municipal buildings and operations);
- the creation of a New York City Panel of Climate Change—consisting of climate change scientists and representatives from legal, insurance, and risk-management firms—tasked with providing information about key climate

BOX 4.2
Analytic Deliberation

The idea of linking science and decision making is explored in a growing body of research literature.[a] One specific approach suggested in this literature is "analytic deliberation," an iterative process in which interested parties initially define objectives and select options to consider, work with experts to generate and interpret relevant new information, and use that information to revise objectives and make choices.[b]

When addressing a problem such as climate change, analytic deliberation processes are particularly valuable because stakeholder discussions help to inform decision makers and the scientific community about local conditions—which is critical because many actions to limit emissions or adapt to climate change must be tailored to local conditions in order to be successful. It also helps to ensure that the scientific community is aware of public concerns and can thus direct research attention to those concerns. Finally, it helps ensure two-way dialogue between scientific experts and the public, which is a more effective communications strategy than a one-way flow of information from scientists to the public. To be successful, however, these deliberative processes require recognizing and overcoming many common obstacles to effective communication. Several NRC and other studies offer guidance on addressing such communication challenges.[c]

One example of this type of engagement process can be found in the NOAA Regional Integrated Sciences and Assessments (RISA) Program,[d] which supports teams at universities and regional centers to conduct research related to climate impacts (e.g., on fisheries, water, wildfire management, agriculture, tourism and recreation, public health, coastal management, infrastructure)—with the goal of helping to inform the decisions of regional-level planners and managers. RISA projects typically involve an array of stakeholders in framing problems for research, and they emphasize collaboration among scientists and decision makers. Although still a relatively new effort, the RISA programs are important test beds for learning how to apply principles of stakeholder engagement for informing decisions about adapting to climate change.

---

[a] See, for instance, D. H. Guston, "Boundary organizations in environmental policy and science: An introduction" (*Science, Technology, and Human Values* 26:399-408, 2001); D. W. Cash, W. C. Clark, F. Alcock, N. M. Dickson, N. Eckley, D. H. Guston, J. Jäger, and. R. B. Mitchell, "Knowledge systems for sustainable development" (*Proceedings of the National Academy of Sciences* 100:8086-8091, 2003); NRC, *Public Participation in Environmental Assessment and Decision Making* (Washington, D.C.: National Academies Press, 2008); NRC, *Informing Effective Decisions*.

[b] See also NRC, *Understanding Risk: Informing Decisions in a Democratic Society*, eds. P. C. Stern and H. Fineberg (Washington, D.C.: National Academy Press, 1996); NRC, *Public Participation*; O. Renn, *Risk Governance: Towards an Integrative Approach* (Geneva, Switzerland: International Risk Governance Council, 2005).

[c] NRC, *Understanding Risk*, *Public Participation*, *Informing Decisions*, *Advancing the Science*, and *Informing Effective Decisions*; EPA, *Improved Science-Based Environmental Stakeholder Processes: A Commentary by the EPA Science Advisory Board*, EPA-SAB-EC-COM-01-006 (Washington, D.C.: U.S. Environmental Protection Agency, 2001).

[d] http://www.climate.noaa.gov/cpo_pa/risa/.

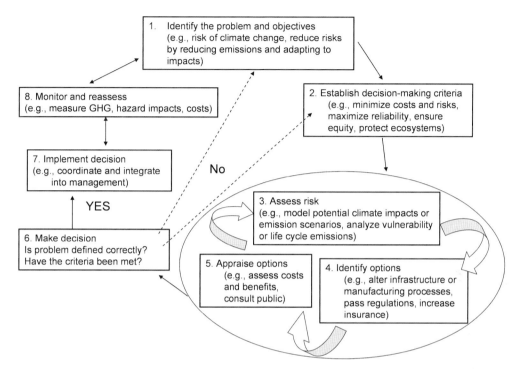

FIGURE 4.1 Illustration of the steps in an iterative risk management approach for addressing climate change. SOURCE: Adapted from R. I. Willows and R. K. Connell, *Climate Adaptation: Risk, Uncertainty, and Decision Making*, UKCIP Technical Report (Oxford, UK: UK Climate Impacts Programme, 2003).

hazards for the city and the surrounding region, likelihoods of their occurrence, and potential implications for critical infrastructure;

- a Climate Change Adaptation Task Force, consisting of over 40 public and private sector stakeholders, that developed a coordinated adaptation plan for the city; and
- a Policy Working Group that identified codes, rules, and regulations governing city infrastructure that may need to be changed or created to help the city cope with climate change.

These activities explicitly call for iterative processes in which goals and strategies are regularly monitored and reassessed, to determine whether intended objectives are being met, to discern any unforeseen consequences, and to allow for periodic corrections. NRC, *Adapting to the Impacts* and *Informing Effective Decisions* contain more details about these New York City activities, and other case studies illustrating how iterative risk management principles are being implemented in both the public and private sectors.

DECISION CRITERIA IN AN ITERATIVE RISK MANAGEMENT FRAMEWORK

This section explores some of the criteria that would be most critical for climate-related decision making in the context of an iterative risk management framework.

Risk reduction potential. A key benefit desired for any action taken to respond to climate change is the potential to actually reduce climate-related risks, by either reducing the likelihood of adverse events (i.e., limiting climate change) or reducing vulnerability to such events (i.e., adapting to climate change) or ideally both. Although risk reduction potential is often difficult to quantify, it can provide a basis for choosing between different options under consideration. As an example, to respond to sea level rise, a community may face a choice between building sea walls to protect buildings and infrastructure or moving those assets to higher ground. The latter option would be more expensive and disruptive in most situations, but it could protect against a broader range of outcomes.

In certain cases, response options can reduce some risks while increasing others, thus requiring trade-offs among risks. For example, promoting more widespread use of air-conditioning to adapt to higher summer temperatures will undermine efforts to limit climate change, to the extent that the additional electricity required is generated by sources that emit GHGs. In other cases, an option may offer complementary risk reduction benefits. For example, improvements in the energy efficiency of buildings and their cooling systems can both constrain the growth of GHG emissions and reduce the threat that heat waves pose to vulnerable populations.

Some actions—such as those involving investment in new technologies, infrastructure, and workforce capacity—may offer little or no direct risk reduction potential themselves but can open the door to future options that may significantly reduce risk. For example, investing in development of a "smart grid" would provide flexibility for integrating distributed renewable electricity generation, and investing in the training of scientists and engineers can improve scientific understanding and the likelihood of significant technological breakthroughs over time.[12] Other options, in contrast, may foreclose future risk-reducing possibilities. For example, continuing to build new coal-fired power plants will lock in further dependence on GHG-intensive energy sources (unless commercial-scale carbon capture and storage soon become widely implemented).

The field of risk analysis, which has a large research literature,[13] offers general guidance on the process of estimating risk reduction potential. For the issue of climate change in particular, the many uncertainties and personal judgments that are inevitably involved in weighing different types of risks have led some analysts to develop methods that synthesize the judgments of many experts.[14]

Feasibility and effectiveness. The potential for any given climate change response action to reduce risk must be measured against the feasibility (which may encompass technical, economic, and political feasibility) and the likely effectiveness of that action. A good deal is known, for example, about the feasibility and effectiveness of certain renewable energy technologies (e.g., wind), while relatively little is known about the feasibility of others (e.g., tidal).[15] Where an option promises substantial risk reduction but has high costs and is of unproven effectiveness, the best response may be investment in further study or pilot testing to reduce unknowns surrounding its application.

Questions about feasibility and effectiveness also apply to policy tools. Insights about the effectiveness of different policy approaches can be gained from the research literature and also from the diverse experience of state and local governments, efforts in other nations, and U.S. federal programs in analogous contexts. For instance, to learn about the effectiveness of cap-and-trade programs, one can look to the experiences of the Regional Greenhouse Gas Initiative of the northeastern states, of the European Union's emission trading system, and of the acid rain cap-and-trade program under Title IV of the Clean Air Act.[16]

Cost and cost-effectiveness. In a world of finite resources, cost and cost-effectiveness are important criteria for helping policy makers decide among different response options. Cost-effectiveness analysis assumes a similar level of risk reduction among options—if two options have similar risk reduction potential and likely effectiveness, a decision maker would choose the option with lower costs. In contrast, cost-benefit analysis is typically used to determine an optimal risk reduction strategy that balances costs and social benefits. As discussed earlier, however, cost-benefit and cost-effectiveness analysis have some important limitations when it comes to analyzing climate choices.

The cost of some options may be so disproportionate to risk reduction potential as to be clearly unreasonable: for instance, certain actions may threaten widespread business closures or other economic impacts that render the option unwise or politically impractical. (For this reason, cost considerations could be viewed as one aspect of the "feasibility" criterion discussed above.) In contrast, some options may be warranted by the positive economic returns or ancillary benefits they offer, even without consideration of climate-related benefits—including, for example, programs to encourage energy efficiency that yield a positive net economic benefit.[17]

Ancillary costs and benefits. Some options designed to reduce climate-related risks may have negative impacts on national interests in other areas, such as ecosystem services, human health, and national security. Examples include nuclear proliferation risks associated with increased reliance on nuclear power, and risks to ecological systems and food security stemming from increased assignment of agricultural land to biofuels production.

Other policies designed to limit or adapt to climate change may have significant ancillary benefits. For example, increasing energy efficiency to limit GHG emissions can also reduce emissions of conventional pollutants,[18] and reducing GHG emissions from the transportation sector could potentially reduce petroleum consumption and thus the nation's vulnerability to high oil prices and oil-supply disruptions.[19] Encouraging carbon sequestration through soil and forest management practices (e.g., minimum tillage practices, reducing timber harvesting, improving manure management, reducing livestock herd size) may also offer the benefits of helping to control nutrient runoff, soil erosion, and habitat loss.[20] It is wise to consider potential co-benefits of this kind when choosing among alternative possible strategies for reducing climate risks.

Equity and fairness. Equity and fairness concerns are important criteria for evaluating any public policy option. International debates have focused on how to fairly allocate the burdens of addressing climate change between developed and developing countries. Intergenerational justice debates center around defining the present generation's obligations to help ensure the well-being of future generations. Domestic policy debates have focused on how policies for reducing GHG emissions may alleviate or exacerbate burdens among different parts of society (e.g., on low-income households or on geographical regions that are heavily dependent on fossil fuel-based industries) and on the socioeconomic distributional impacts of actions taken to adapt to climate change.

Consider, for example, the case of lower-income households, which consume less energy per capita and thus contribute proportionately less to GHG emissions relative to more affluent households. Energy purchases are a larger fraction of their total consumption, and therefore they are more affected by changes in energy prices.[21] Limited discretionary income may also preclude lower-income households from participating in energy efficiency initiatives that would reduce their energy costs over the longer term. At the same time, lower-income households may suffer disproportionately from the impacts of climate change.[22] Some ways in which policy design can help address such concerns are discussed in Chapter 5.

International considerations. America's climate choices affect and are affected by the global dimensions of climate change. U.S. emissions reductions alone will not be adequate to avert dangerous climate change risks; rather, our emission reductions must be accompanied by comparable actions from all other major emitters. U.S. climate policies can potentially have a major effect on the actions other countries take, and this potential represents another important criterion for evaluating domestic response options. In general, domestic policies that help leverage broader international-scale efforts (for example, cooperative research and development programs in clean energy technology) can be expected to reduce overall climate risk more than policies that affect U.S. emissions alone. Similarly, in comparing the advantages and disadvantages of different policy options for reducing U.S. GHG emissions (e.g., cap-and-trade programs, carbon taxes, regulatory approaches), each should be considered in the context of how they link domestic policies to global efforts.

Robustness. Given the uncertainties inherent in predicting future climate change and its impacts, as well as the difficulty of predicting technological, social, and economic developments, there is a great strategic advantage in pursuing response options that can perform well under a wide range of possible futures. For instance, sound risk management in the agricultural sector may include investing in the development of crop varieties that are resilient to a wide range of temperature and precipitation conditions. As another example, market-based regimes offer an advantage over industry-specific performance standards because the former approach has a higher likelihood of continued effectiveness under varying future economic or technological conditions.[23]

When the likelihood of different future outcomes is not well known, pursuing multiple options (i.e., a portfolio approach) and other "hedging" strategies can help ensure a robust response. For example, it would be prudent to invest in multiple new energy technologies to meet future needs because the ultimate success of any one new technology is always uncertain. As another example, it is prudent to design the infrastructure for transportation, water, and utilities to withstand a range of weather extremes including intense rainfall, flooding, and drought scenarios. Ensuring robustness may also include strengthening general adaptive capacity through early warning systems and disaster response preparations.

The degree to which any particular policy option meets the different criteria listed above depends not only on the type of policy but also its scope and stringency. For example, an overly weak auto fuel efficiency standard may be cheap and politically feasible but not very effective in reducing climate-related risks, whereas an overly tough standard may promise high levels of risk reduction but be very expensive, pose significant equity concerns, and be difficult to implement successfully.

Ultimately, any choice involves weighing multiple criteria. Decision makers will differ in their judgments about which criteria are most important and in their methods for dealing with uncertainties. Even when it is possible to characterize how different response actions rank under the different criteria, this information may not necessarily point to a preferred action or strategy. Rather, this information provides a basis on which decision makers can make reasoned judgments and engage in informed debates. The decision sciences offer a variety of methods for helping decision makers evaluate and make trade-offs among options,[24] but even these methods do not obviate the need for deliberation and judgment.

CHAPTER CONCLUSION

In the committee's judgment, iterative risk management—which emphasizes taking action now, but in doing so, being ready to learn from experience and adjust these efforts later on—offers the most useful approach for guiding America's climate choices. The successful application of this approach requires broad-based continuous learning by the scientific community together with decision makers in the government, the private sector, and the general public.

CHAPTER FIVE

Key Elements of America's Climate Choices

A comprehensive and effective response to climate change requires a diverse portfolio of actions that evolve with new information and understanding.

As discussed in Chapter 4, a balanced risk management strategy for addressing climate change requires an integrated portfolio of policy options, including actions aimed at reducing the likelihood of adverse outcomes and actions aimed at reducing the damages such outcomes could cause. It further requires making investments over time to advance the knowledge on which future decisions will be based, to expand the options available to decision makers in the future, and to ensure that decision makers at all levels (including the public) have the information necessary to make decisions that properly reflect new knowledge and new options. Cutting across all of these elements are needs for international engagement and for coordinating the different actors and elements of an overall response strategy. This chapter offers recommendations along each of these dimensions, with an emphasis on near-term responses that enhance the capacity for, and reduce the costs of, more substantial responses that may be chosen in the more distant future.

LIMITING THE MAGNITUDE OF CLIMATE CHANGE

Limiting the magnitude of climate change requires stabilizing atmospheric greenhouse gas (GHG) concentrations, which in turn requires reducing emissions of these gases so that their emissions are no greater than the rate at which they are naturally removed from the atmosphere. The basic opportunities available for reducing GHG emissions include restricting or modifying activities that release GHGs (e.g., burning of fossil fuels), removing CO_2 from the waste stream of large point sources of emissions and sequestering it underground (carbon capture and storage), or augmenting natural processes that remove GHGs from the atmosphere, for example by managing agricultural soils or forests to increase the rate at which they sequester carbon (*post-emission GHG management*). In addition, GHG emissions could potentially be offset by enhancing the reflection of solar radiation back to space (*solar radiation management*), which together with some post-emission GHG management strategies is sometimes referred to as geoengineering (see Box 5.1). NRC, *Advancing the Science* and *Limiting the Magnitude* review the technologies and practices that are available for pursuing these various opportunities. Here we provide a very brief overview, starting with the more general issue of setting goals for limiting the magnitude of climate change.

BOX 5.1
Geoengineering

Geoengineering, applied to climate change, refers to deliberate, large-scale manipulations of the Earth's environment intended to offset some of the harmful consequences of GHG emissions, and it encompasses two very different types of strategies: solar radiation management and post-emission GHG management.[a] Many proposed geoengineering approaches are ambitious concepts with global environmental consequences; as such, they have attracted a great deal of attention. In general, however, current scientific knowledge of the efficacy and overall risk reduction potential of most geoengineering approaches is limited.

Solar radiation management (SRM) involves increasing the reflection of incoming solar radiation back into space. Some SRM approaches can, in theory at least, produce substantial cooling quickly and thus could potentially be used in the case of "climate emergencies" involving unexpectedly rapid warming or severe impacts. A much-discussed example is the proposal to continuously inject large quantities of small reflective particles (aerosols) into the stratosphere. This would mimic some effects of sustained large volcanic eruptions, which have been observed to cool the earth's surface measurably for months. Another SRM strategy sometimes proposed is to increase the reflectivity of the Earth's surface through widespread use of "white roofs."

The potential benefits of many SRM strategies are offset by potential risks. In the case of aerosol injection strategies, for example, significant regional or global effects on precipitation patterns could occur,[b] potentially placing food and water supplies at risk. SRM alone would also do nothing to slow ocean acidification, since CO_2 concentrations in the atmosphere and ocean would continue to rise. Thus, it is unclear if any of the proposed large-scale SRM strategies could actually reduce the overall risk associated with human-induced climate change.[c]

Large-scale proposals for post-emission GHG management generally involve either removing CO_2 from the atmosphere by direct air capture technologies or managing ecosystems on land or in the ocean to increase their natural uptake and storage of carbon. Strategies for enhancing carbon sequestration in soils and forests (which are often viewed as standard strategies for limiting climate change rather than geoengineering) are relatively well understood and offer important opportunities for reducing net GHG emissions in some parts of the world. However, changes in the sequestration rates of these systems are often difficult to quantify, and the potential effectiveness of these strategies may decline over time due to saturation effects.[d]

A more controversial post-emission GHG management strategy which has actually been tested at small scales is fertilizing the ocean by adding iron to iron-poor waters (or adding other limiting nutrients or minerals) to increase removal of CO_2 from the atmosphere by phytoplankton. Concerns have been raised both about the efficacy of this approach and about possible risks that it might pose to marine

---

Setting Goals

Concrete, quantitative goals for limiting the magnitude of climate change offer the benefit of allowing all parties involved to have a common sense of purpose and a clear metric against which to measure progress. Finding agreement on quantitative

ecosystems.[e] Other post-emission approaches involve removing CO_2 from the air though chemical processes, but as with conventional carbon capture and storage from point emission sources, direct air capture schemes require reliable geological repositories for the removed CO_2. In general, most CO_2 removal approaches seem to pose fewer ancillary risks than SRM approaches, but they appear likely to be expensive, and because they would have only a gradual effect on atmospheric GHG concentrations, they would not have the potential to produce substantial cooling quickly.

Because some forms of geoengineering would have consequences that span national boundaries, international legal frameworks are needed to govern the development and possible deployment of these options. Such frameworks need to include a clear definition of the "climate emergency" that would trigger deployment of large-scale SRM, and criteria for whether, when, and how SRM (and some versions of post-emission GHG management) would be tested —recognizing that even the act of field testing may create international tensions. More fundamentally, intentional alteration of the Earth's environment via geoengineering raises significant ethical issues, including the distribution of risks among population groups in both present and future generations, as well as challenging questions of public perceptions and acceptability.[f]

In conclusion, geoengineering approaches may conceivably have a role to play in future climate risk management strategies, particularly if efforts to reduce global GHG emissions are unsuccessful or if the impacts of climate change are unexpectedly severe. At present however, the costs, benefits, and risks of many geoengineering approaches are not well understood. In the committee's judgment, it would therefore be imprudent to use certain geoengineering approaches (in particular, SRM and ocean fertilization strategies) to manipulate the Earth's environment in the near future, and it would be unwise to assume they will be attractive options even in the more distant future. We recommend instead a program of research to better understand the potential effects of different geoengineering options and efforts to address the international governance issues raised by many geoengineering proposals.

[a] See, e.g., Royal Society, *Geoengineering the Climate: Science, Governance, and Uncertainty*, RS policy document 10/09 (London: The Royal Society, 2009); American Geophysical Union, *Geoengineering the Climate System. A Position Statement of the American Geophysical Union* (Adopted by the AGU Council on 13 December 2009); American Meteorological Society, *Geoengineering the Climate System. A Policy Statement of the American Meteorological Society* (adopted by the AMS Council on 20 July 2009).
[b] G. C. Hegerl and S. Solomon, "Risks of climate engineering" (*Science* 325[5943]:955-956, 2009, doi: 10.1126/science.1178530).
[c] See NRC, *Advancing the Science*, Chapter 15 for additional discussion of proposed SRM approaches, including the research needed to better understand their potential efficacy and risks.
[d] See NRC, *Advancing the Science* and *Limiting the Magnitude* for further discussion and references.
[e] K. O. Buesseler, S. C. Doney, D. M. Karl, P. W. Boyd, K. Caldeira, F. Chai, K. H. Coale, H. J. W. De Baar, P. G. Falkowski, K. S. Johnson, R. S. Lampitt, A. F. Michaels, S. W. A. Naqvi, V. Smetacek, S. Takeda, and A. J. Watson, "Environment: Ocean iron fertilization—Moving forward in a sea of uncertainty" (*Science* 319[5860]:162, 2008).
[f] NRC, *Advancing the Science*.

targets is, however, an inevitably contentious process, and a failure to reach consensus on such targets can become a barrier to moving ahead with meaningful actions. It is of course possible to proceed with meaningful actions to limit GHG emissions in the absence of universally-accepted quantitative goals. Nonetheless, it is important to understand the different types of goals that are being actively debated at national and international levels.

At the international level, a commonly discussed goal is the tolerable increase in global average surface temperature relative to pre-industrial times. (The goal of limiting global average temperature rise to 2°C (3.6°F) has been agreed to in a number of major international platforms,[1] although there is ongoing scientific debate about whether that actually represents a "safe" threshold for limiting climate change.[2]) For any given global temperature goal, corresponding goals can then be derived for atmospheric GHG concentrations that would give a reasonable chance of meeting the temperature goal, for global GHG emissions limits that would give a reasonable chance of meeting those GHG concentration goals, and, finally, for national GHG emission limits that would collectively achieve the needed global emission reductions. These relationships are complicated, however, by a variety of scientific uncertainties and value judgments (see Figure 5.1).[3]

A global mean temperature limit is not a goal that can be directly controlled, but rather, is an emergent property of the decisions made by countless governments, private sector actors, and individuals around the world, and of the earth system processes that determine how emissions affect the earth's climate. Operationally, domestic-level response strategies require metrics that can be directly tracked and controlled at the national level. For the U.S. national goal, the America's Climate Choices (ACC) panel report *Limiting the Magnitude of Future Climate Change* recommends setting a "budget" for cumulative domestic GHG emissions over a set period of time—a recommendation the committee supports. The budget concept has also been proposed in the context of global emissions.[4]

It is beyond the mandate of this committee to recommend specific global or national emission budget goals because such goals are based in large part on value judgments about what is an acceptable degree of risk, and what is a fair U.S. share of the global emissions-reduction burden. Nor do we try to evaluate the risks of adverse climate impacts associated with different possible U.S. emission goals, because such risks ultimately depend on global emissions, not U.S. emissions alone. We do suggest, however, that in the context of iterative risk management, any such goals need to be periodically revisited and revised over time, in response to new information and understanding.

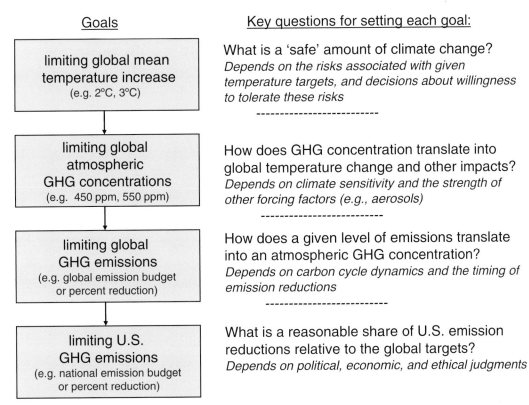

Goals

Key questions for setting each goal:

| limiting global mean temperature increase (e.g. 2°C, 3°C) | **What is a 'safe' amount of climate change?** *Depends on the risks associated with given temperature targets, and decisions about willingness to tolerate these risks* |

| limiting global atmospheric GHG concentrations (e.g. 450 ppm, 550 ppm) | **How does GHG concentration translate into global temperature change and other impacts?** *Depends on climate sensitivity and the strength of other forcing factors (e.g., aerosols)* |

| limiting global GHG emissions (e.g. global emission budget or percent reduction) | **How does a given level of emissions translate into an atmospheric GHG concentration?** *Depends on carbon cycle dynamics and the timing of emission reductions* |

| limiting U.S. GHG emissions (e.g. national emission budget or percent reduction) | **What is a reasonable share of U.S. emission reductions relative to the global targets?** *Depends on political, economic, and ethical judgments* |

FIGURE 5.1 A schematic illustration of the steps involved in setting goals for limiting the magnitude of future climate change, and some key questions and uncertainties that need to be considered in each of these steps.

Reducing Global Emissions

The United States currently accounts for roughly 20 percent of global CO_2 emissions, despite having less than 5 percent of the world's population. The U.S. percentage of total global emissions is projected to decline over the coming decades, however, mainly because emissions from rapidly developing nations such as China and India will continue to grow (see Figure 3.1). International engagement challenges are discussed later in this chapter, but it is worth emphasizing here the central point that poorer nations usually find requests by the United States to limit their emissions unjustified for several reasons: because current per-capita emissions and standard of living in the United States and other developed nations are far above theirs, because the United States is responsible for the largest share of the historical increase in atmospheric GHG concentrations, and because the United States has not yet been willing to enact its own national policies to limit emissions. As a result of such dynamics, the international community has not yet forged an agreement in which both developed countries and all large, rapidly developing countries commit to binding GHG emission reductions. Forging a comprehensive international agreement will be difficult and possibly infeasible without credible U.S. leadership, demonstrated through strong domestic actions.[5]

In addition to this need for demonstrating leadership through strong domestic actions, America's climate change response strategies need to include cooperative international efforts aimed at helping developing countries advance their economies along less carbon-intensive pathways than were followed by today's industrialized nations. This is primarily because reducing global GHG emissions requires limiting the growth in emissions from developing countries. Additional motivation comes from the fact that it is generally less expensive to reduce emissions in developing nations than in developed ones (although the evidence can vary considerably depending on the specific context),[6] and because developing nations often present more significant opportunities for ancillary benefits such as reducing local air pollution.

Emission Offsets

Offsets can be used at either the domestic or international level to help lower the cost of reducing emissions. In most cases, an offset system allows actions that remove or prevent GHG emissions in one place to cancel (or offset) an equivalent amount of emissions elsewhere. Offsets can include investments in agriculture and land management, reforestation, energy efficiency, capture or destruction of industrial gases and methane, or low carbon energy generation such as renewables. International offset programs such as the Clean Development Mechanism (CDM), that allow GHG emitters in the United States or other developed countries to pay for emission reductions in developing countries, can also be a potentially important mechanism for engaging those countries in emission-reduction efforts. Although the United States does not participate in the CDM, it has been involved in discussion of other international mechanisms similar to offsets, particularly proposals to reduce emissions from deforestation and forest degradation (often referred to as REDD).

The use of offsets can be complex and fraught with pitfalls, however. Some offsets, such as capturing methane from livestock waste, are relatively straightforward to quantify and implement. Others, such as sequestering carbon in soils and forests, or those where investments are made in small scale technologies such as improved wood stoves, present many challenges. Quantifying the size of the offset requires not only knowing how much carbon is being sequestered by the system (which in turn requires accurate baselines and monitoring) but also knowing what would have happened in the absence of the offset program—i.e., offsets only contribute to reducing emissions if they are clearly *additional* to programs that would have been implemented anyway. In contrast, a cap-and-trade or carbon tax system (that does not allow use of offsets) considers only what actually is emitted; it does not require estimates, often controversial, of what would have been emitted absent the policy or action considered.

Offsets also raise concerns about *emissions leakage*: for example, if the demand for timber is unchanged (and there are no global emission caps), saving one forest that would have been cut could simply increase the pressure to cut other forests that would otherwise have been spared. A similar problem occurs with the so-called *rebound* effect, when savings from energy efficiency actions are invested in other activities that produce GHGs. Another complication arises from the fact that the amount of sequestered carbon can change over time. For instance, if a forest grown to offset carbon emissions from elsewhere burns down 10 years later, the emissions reductions provided by the offset will be lost.

Finally, the ancillary ecological and social impacts of offset programs can be either positive or negative, depending on whether they are guided by sound sustainable development or land management principles and practices, including respect for local property rights.[7] For these reasons, the inclusion of offsets as a major component of U.S. climate policy will require rigorous rules, standards, and accounting procedures to ensure claimed emissions reductions are real and sustained.

Reducing U.S. Emissions

The nation's efforts to reduce GHG emissions depend to a large degree on private sector investments (in areas such as technology development, physical assets, manufacturing operations, and marketing and delivery of goods and services) and on the behavioral and consumer choices of individual households. But federal, state, and local governments have a large role to play in influencing these key stakeholders through effective policies and incentives. In general, there are four major tool chests from which to select policies for driving GHG emission reductions:

- pricing of emissions by means of a tax or cap-and-trade system;
- mandates or regulations, which includes full-scale programs of controls on emitters (for example through the Clean Air Act) and more narrowly targeted mandates such as automobile fuel economy standards, appliance efficiency

standards, labeling requirements, building codes, and renewable or low-carbon portfolio standards for electric generation;[8]
- public subsidies through the tax code, appropriations, or loan guarantees; and
- providing information and education and promoting voluntary measures.[9]

A comprehensive national program would likely use tools from all four of these areas. Most economists and policy analysts have concluded, however, that putting a price on CO_2 emissions (that is, implementing a "carbon price") that rises over time is the least costly path to significantly reduce emissions and the most efficient means to provide continuous incentives for innovation and for the long-term investments necessary to develop and deploy new low-carbon technologies and infrastructure.[10] A carbon price designed to minimize costs could be imposed either as a comprehensive carbon tax with no loopholes or as a comprehensive cap-and-trade system that covers all major emissions sources. (Pricing systems that are not comprehensive can also produce substantial reductions, though at higher per-ton costs.) Both of these could be effective tools; however, cap-and-trade policy offers the advantage of specifying emissions goals. Moreover, if several nations have cap-and-trade systems and international trading is permitted, firms in rich nations can reduce their costs—and total global costs—by paying for less expensive emissions reductions in other nations, rather than by making expensive reductions themselves.[11]

Meeting stringent national emission-reduction goals also requires the carbon price to rise to levels that are high enough to ensure the necessary investments are made in energy-efficient buildings and equipment, low-carbon energy production technologies, and other key areas, especially over the long run as stocks of equipment and infrastructure turn over. Estimating possible future carbon prices, which depends on many unpredictable factors, such as the pace of technology development, is beyond the scope of this study; but NRC, *Limiting the Magnitude* does contain a detailed discussion of future carbon price projections made in the recent multi-model studies of the Energy Modeling Forum.[12]

In addition to a price on carbon, there is a need for complementary policy measures that help to overcome market failures not fully addressed by a carbon price.[13] Complementary policies may also be needed to overcome institutional barriers that inhibit responses to carbon prices and/or slow the penetration of new low-carbon technologies.[14] Examples of such barriers include outdated building codes and regulatory systems[15] and the information-related problems that reduce incentives for builders and home owners to invest in energy-efficient homes and appliances.[16] Complementary policies must be chosen strategically, however—an optimal policy reduces emissions where it is cheapest to do so, not taking all possible measures, nor requiring all sectors of the economy to participate equally. *Limiting the Magnitude* examines the types of complementary policies that are most useful for ensuring rapid progress in key areas such as household-level energy efficiency, development and use of renewable energy technologies, and retiring/retrofitting existing emissions-intensive equipment and infrastructure.

As a matter of political reality, a comprehensive carbon-pricing strategy of the sort described above may not be feasible in the near term. A strategy relying solely on other types of policies would involve higher costs but would still encourage near-term emission reductions and thus reduce the need to make costly reductions later. These policies range from relatively simple measures such as supporting R&D on low-carbon technologies and reducing behavioral and institutional barriers to energy efficiency, to more ambitious steps such as a nationwide renewable portfolio standard or a cap-and-trade system covering only electric power plants.[17] To minimize the long-run costs of reducing emissions, however, it is important to avoid policies that may make it more difficult later (either economically or politically) to adopt a comprehensive carbon-pricing policy. This includes, for instance, policies that would implicitly or explicitly exempt some sources from a subsequent carbon tax or a broader emissions cap. It may also be necessary to avoid certain policies that have unacceptable equity and competitiveness impacts (see Box 5.2).

At the time of writing this report, the EPA is in the process of promulgating new rules to constrain CO_2 emissions using the current authorities of the Clean Air Act. These rules, if adopted,[18] will likely achieve emission reductions and may also stimulate innovation, but the regulatory strategy is not as likely as a well-crafted pricing strategy to provide continuous incentives to find the cheapest path to significant GHG reductions.[19]

RECOMMENDATION 1: In order to minimize the risks of climate change and its adverse impacts, the nation should reduce greenhouse gas emissions substantially over the coming decades. The exact magnitude and speed of emissions reduction depends on societal judgments about how much risk is acceptable. However, given the inertia of the energy system and long lifetime associated with most infrastructure for energy production and use, it is the committee's judgment that the most effective strategy is to begin ramping down emissions as soon as possible.

Emission reductions can be achieved in part through expanding current local, state, and regional level efforts, but analyses suggest that the best way to amplify and accelerate such efforts, and to minimize overall costs (for any given national emissions

---

BOX 5.2
Equity and Competitiveness Issues

Significantly reducing U.S. GHG emissions, however it is accomplished, will produce "winners" and "losers" along several dimensions. Increasing the price of carbon-intensive energy, for instance, will have a disproportionate impact on those who need to drive long distances to work and residents of some coal-mining communities. Basic notions of fairness require that adverse energy price impacts on those least able to bear them be identified and addressed. Carbon-related revenues, obtained from carbon taxes or auctioning of emissions allowances in a cap-and-trade system, would provide resources that could be used for this purpose. Alternative or additional policy measures that make incentive-based climate change policies more accessible to low-income households (e.g., graduated subsidies or tax credits for home insulation improvements) may also be appropriate. Directly engaging economically disadvantaged and other vulnerable communities in the policy planning process helps allow the legitimate interests of those communities to be addressed, while nonetheless allowing broadly desirable investments to be made.

In an economy with substantial unemployment, expansion of labor-intensive activities like retrofitting buildings for increased energy efficiency can be an attractive option for increasing job opportunities.[a] A transition to a low-carbon economy would inevitably produce gains in some sectors and occupations and losses in others, and some studies suggest that such a transition will probably have only a small net impact on the overall level of U.S. employment.[b] For those sectors and regions that are at greatest risk of job losses, this transition could be smoothed through targeted support for education and training programs.

---

reduction target), is with a comprehensive, nationally uniform, increasing price on CO_2 emissions, with a price trajectory sufficient to drive major investments in energy efficiency and low-carbon technologies. In addition, strategically-targeted complementary policies are needed to ensure progress in key areas of opportunity where market failures and institutional barriers can limit the effectiveness of a carbon pricing system.

If a pricing strategy proves to be politically infeasible, second-best approaches may include the expansion of regional, state, and local initiatives already under way, along with the adoption of national-level mandates or performance standards, some of which could potentially be implemented through the Clean Air Act. The committee suggests that new mandates and standards leave as much flexibility as possible for the private sector to choose the means necessary (i.e., the technological options) for meeting stated emission-reduction goals and leave room for later adoption of a pricing strategy.

Finally, de-carbonizing the U.S. energy system—or failing to do so—could have a significant impact on the competitiveness of some U.S. industries. The European Union (EU) has already increased its reliance on renewable energy and put a price on CO_2 emissions from major sources, without detectable adverse economic effects.[c] China has now placed low carbon and clean energy industries at the heart of the country's strategy for industrial growth, and is making large scale public investments (for instance, in "smart grid" energy transmission systems) to support this growth.[d] If others continue to press in this direction but the United States does not, firms operating in the United States could find themselves increasingly out of step with the rest of the world, and without the robust domestic markets for climate-friendly products that their competitors in the EU and elsewhere would enjoy. Moreover, U.S. firms in energy-intensive sectors could be disadvantaged relative to their more energy-efficient foreign competitors if energy prices rise in coming decades (as many observers expect) regardless of whether global actions are taken to reduce GHG emissions. Firms operating in the United States might also face tariffs on their exports to countries that have emissions caps in place and are seeking to protect their industries from the competition posed by countries without such caps.

---

[a] R. Pollin, H. Garrett-Peltier, J. Heintz, and H. Scharber, *Green Recovery: A Program to Create Good Jobs and Start Building a Low-Carbon Economy* (Washington, D.C.: Center for American Progress, 2008)

[b] NRC, *Limiting the Magnitude*; CBO, *The Economic Effects of Legislation to Reduce Greenhouse Gas Emissions* (Washington, D.C.: Congressional Budget Office, 2009); H. Huntington, *Creating Jobs with Green Power Sources*, Energy Modeling Forum OP64, (Stanford, CA: Stanford University, 2009).

[c] A. D. Ellerman, F. J. Convery, and C. de Perthuis, *Pricing Carbon: The European Union Emissions Trading Scheme* (Cambridge, UK: Cambridge University Press, 2010).

[d] *http://www.energychinaforum.com/news/42628.shtml* (accessed Feb. 20, 2011).

Finally, as with all elements of an iterative risk management strategy, actions taken to reduce GHG emissions need to be carefully monitored. Decisions made today (e.g., regarding emission targets, price schedules, sectors chosen for special attention) will require periodic reevaluation in light of new developments in climate science, in technological capabilities, in costs, and in understanding the impacts of response policies themselves (for instance, understanding how a carbon pricing system implemented in a less than comprehensive form will actually influence investments). In this regard, long-term emissions goals that stretch out for decades are useful and probably necessary but would likely need to be revisited over time. This need for the capacity to adjust polices in response to new information and understanding must be balanced against the need for policies to be sufficiently durable and consistent to attract substantial investment and encourage long-term changes in behavior. There is a natural tension between these goals and designing mechanisms to provide both durability and flexibility poses a key challenge for climate change governance.

REDUCING VULNERABILITY TO CLIMATE CHANGE IMPACTS

Even if substantial global GHG emissions reductions are achieved, human societies will need to adapt to some degree of climate change. In the committee's judgment, choosing to do nothing now and simply adapt to climate change as it occurs, perhaps by accepting losses, would be an imprudent choice for a number of reasons. Even moderate climate change will be associated with a wide range of impacts on both human and natural systems (see Chapter 2), and the possibility of severe climate change with a host of adverse outcomes cannot be ruled out (especially if GHG emissions continue unabated). Moreover, as climate change progresses, unforeseen events may pose serious adaptation challenges for regions, sectors, and groups that may not now seem particularly vulnerable. Thus, just as the committee recommends that America embark on a course of substantial emission reductions, we recommend that America take proactive actions to mobilize the nation's capacity to adapt to future climate changes. This dual-path strategy will reduce the risks of future climate-related damages more than pursuing either path alone.

Mobilizing Now for Adaptation

As discussed in NRC, *Adapting to the Impacts,* proactive adaptation involves preparing for the impacts of projected local and regional changes in climate before they occur. Examples of such changes (also described in Chapter 2 and references therein) include reduced surface water supply in America's rapidly growing western regions and increased vulnerability of the Gulf Coast to sea level rise, especially in low-lying areas that are already subject to land subsidence and other environmental changes. Global climate models can provide robust projections of changes in some regions (such as reduced precipitation across the southwestern United States), and there are a variety of efforts under way to "downscale" this information to local and regional levels. Currently, however, models are limited in their ability to reliably project many of the local and regional scales that are critical for adaptation decisions; and this situation is only expected to improve gradually as computer power and scientific understanding advance.[20] Regardless, efforts to assess adaptation needs at these scales can help indicate possible risks and key vulnerabilities.

Many types of decisions would benefit from improved insight into adaptation needs and options, but most notable are decisions with long time horizons. These include, for instance, decisions about siting of facilities, conserving natural areas, managing water resources, and developing coastal zones. In these realms, adaptation to climate change is less about doing different things than about doing familiar things differently. For example, throughout history, governments have typically designed and managed water supply systems based (either implicitly or explicitly) on the assumption that future climate conditions will be similar to those experienced in the past. Proactive adaptation to climate change will require these decisions to account for the strong possibility that future climate conditions will be unlike the past (see Box 5.3). Many investments in adaptation will be largely inseparable from routine investments in infrastructure development and upgrading.

Because specific local or regional climate changes often cannot be precisely predicted, long-lived investments whose value may be affected by future climate change are inherently risky. Decisions about such investments accordingly become exercises in risk management. When climate models make differing predictions for a given locality, for example, it is prudent to seek robust options that will lead to acceptable outcomes across the full range of plausible future climate change scenarios (including low-probability but high-risk scenarios). An iterative risk management approach to adaptation also requires effective monitoring and assessment of both emerging impacts and the effectiveness of adaptation actions.

In many cases, adaptation options are available now that both help to manage longer-term risks of climate change impacts and also offer nearer-term co-benefits for other development and environmental management goals (see NRC, *Adapting to the Impacts* for examples). Experience to date indicates that if adaptation planning is pursued collaboratively among local governments, private-sector firms, nongovernmental organizations, and community groups, it often catalyzes broader thinking about alternative futures, which is itself a considerable co-benefit.

An Effective National Adaptation Strategy

Much of the work of adaptation will be done by state, local, and tribal[21] governments, private-sector firms, nongovernmental organizations, and representatives of especially vulnerable regions, sectors, or groups. Decision makers at these levels often lack the resources necessary to perform this work effectively or lack experience in accessing and using information that may be available to inform their decisions. Dealing with vulnerabilities that cross geographic, sectoral, or other boundaries are particularly challenging. Although early adaptation planning is beginning to emerge from the bottom up in the United States, it is hampered by a lack of both knowledge and resources.[22]

The federal government can play a valuable leadership and coordination role for adaptation. In the near term, this includes initiating the development of a *national* (i.e.,

---

BOX 5.3
Adapting to Changing Precipitation Patterns

Different regions of the United States face different types of risks from climate change, thus requiring very different adaptation strategies. Responses to the potential impacts of changing precipitation patterns provide illustrations of these challenges. Climate models generally predict that across the United States, precipitation will increase in northern regions and decrease in the southern and western states. Below are two examples of the different adaptation challenges posed by such changes.

Vulnerability of Mass Transit in New York City to Extreme Precipitation Events. New York City experiences substantial precipitation in all months of the year. Mean annual precipitation has increased slightly over the course of the past century, and inter-annual variability has become more pronounced. Climate models project that in the coming decades, the city will experience an increasing number of heavy rainfall events (for example, the number of days per year where rainfall exceeds 1 inch, 2 inches, or 4 inches).[a] Many components and facilities of rail systems in New York City, such as public entrances and exits, ventilation facilities, and manholes, were built at low elevations. For example, in upper Manhattan parts of the subway system are as much as 180 feet below sea level. Large sections of the system are thus vulnerable to flooding from such heavy precipitation events as well as from sea level rise.[b]

On August 8, 2007, heavy precipitation from a major storm resulted in a system-wide outage of New York City subways during the morning rush hour. Before the system could re-open, eight tons of debris had to be removed and a variety of equipment had to be repaired or replaced.[c] More frequent events like this one can be expected to increase the frequency of transit interruptions unless proactive adaptation steps are taken. The Metropolitan Transit Authority has subsequently invested significantly in the distribution of large pumps throughout the system to reduce sensitivity to extreme precipitation, and the city is also planning investment strategies to reduce exposure.

---

not just federal) adaptation strategy that engages a broad range of decision makers and stakeholders. An important initial step in developing such a strategy is to identify key vulnerabilities to plausible climate changes, which will vary substantially from place to place and among parties within each place. Basic notions of fairness suggest that special near-term consideration be given to identifying and developing adaption strategies for especially vulnerable populations.

Another important federal role is to provide key resources (e.g., scenarios, visualization tools, methods, data) to support vulnerability analyses. There is currently no widely-accepted approach for conducting vulnerability assessments, and much of the data and scientific infrastructure needed to make these assessments robust are lacking. Yet another important component of a national adaptation strategy is to evaluate existing

Projected Changes in Precipitation and Runoff in the Southwest. As a result of climate change, it is projected that the southwestern states will face an overall reduction in average annual precipitation, with corresponding reductions in runoff in the Upper and Lower Colorado River. Recent estimates suggest a reduction of roughly 6 percent for each degree of increase in global mean temperature over the coming century or beyond.[d] The southwest region has long struggled with issues of water availability as population has grown,[e] and the challenges of coping with scarce water resources in this region will only be exacerbated by declines in future water availability due to climate change. A study of potential climate change impacts on the Colorado River system (a system that roughly 30 million people depend upon for drinking and irrigation water) finds that climate changes occurring over the next several decades would increase the risk of fully depleting water reservoir storage far more than the risk expected from population pressures alone. A scenario of 20 percent reduction in the annual Colorado River flow due to climate change results in a near tenfold increase in the probability of annual reservoir depletion by 2057 —a huge water management challenge.[f]

---

[a] New York City Panel on Climate Change, *Climate Risk Information* (New York: New York City Panel on Climate Change, 2009).

[b] NYCSubway.org, June 24, 2005, *http://www.nycsubway.org/perl/stations?207:2659* (accessed July 15, 2009).

[c] Metropolitan Transportation Authority (MTA), *August 8, 2007 Storm Report*. New York: Metropolitan Transportation Authority, 2007, p. 34.

[d] NRC, *Stabilization Targets*, Table 5.3.

[e] See, e.g., NRC, *Colorado River Basin Water Management: Evaluating and Adjusting to Hydroclimatic Variability* (Washington, D.C.: National Academies Press, 2007).

[f] B. Rajagopalan, K. Nowak, J. Prairie, M. Hoerling, B. Harding, J. Barsugli, A. Ray, and B. Udall, "Water supply risk on the Colorado River: Can management mitigate?" (*Water Resources Research* 45:W08201, 2009, doi:10.1029/2008WR00765).

state and federal policies (many involving infrastructure and land use) in light of current knowledge about projected future changes in climate.[23]

Effective federal government leadership means ensuring that federal programs, activities, and planning take climate change into account and, in particular, that maladaptive policies and practices be identified and reformed. This includes, for instance, revising the current definition of the 100-year floodplain.[24] The federal government will also need to take steps to ensure that the national adaptation strategy will be implemented effectively and revised in light of new knowledge. A variety of decision-making processes at all levels will need to be redesigned to ensure that the latest information regarding future climate change is taken properly into account going forward. The exchange of information at the state, local, and tribal levels and between the public and private sectors may be particularly valuable. It will also be important to institutionalize coordination among the many federal agencies with resources and authorities relevant to adaptation and to implement durable research and training programs aimed at enhancing the capability of nonfederal governments and the private sector to adapt to future climate change.

Internationally, the impacts of climate change will disproportionately be felt in those developing nations that lack both the necessary expertise and financial resources for critical investments.[25] The potentially destabilizing impacts of climate change in the developing world have been identified by military and security analysts as a national security issue for the United States. Such impacts also pose a humanitarian concern should climate change, for example, lead to increases in the occurrence or severity of natural disasters. There are also economic implications should climate change affect consumers of key exports or regions of critical food production and other imports. It would thus be prudent for the federal government to support efforts to enhance adaptive capability in the developing world, as well as within the United States. These and other international concerns are further discussed later in the chapter.

Like the challenge of reducing GHG emissions, the challenge of adaptation will be with America for decades. Accordingly, federal efforts need to include not just formulating an initial national adaptation strategy but also creating durable institutions (or in some cases, strengthening existing institutions) that can revise and improve that strategy over time, in light of new knowledge and new policy options.

Regardless of whether the federal government plays this leadership and coordination role in the near term, it would be prudent for nonfederal government leaders, the private sector, and nongovernmental organizations to take proactive steps to reduce vulnerability to climate change, through for instance, vulnerability assessment and investments in preparedness and disaster response. In addition, "horizontal" coordination among subfederal governments and the private sector can help facilitate the efforts of all participants. There is a role as well for nongovernmental organizations that provide expertise and help catalyze the needed coordination.

RECOMMENDATION 2: Adaptation planning and implementation should be initiated at all levels of society. The federal government, in collaboration with other levels of government and with other stakeholders, should immediately undertake the development of a national adaptation strategy and build durable institutions to implement that strategy and improve it over time.

INVESTING TO EXPAND OPTIONS AND IMPROVE CHOICES

A sound strategy for managing climate-related risks requires sustained investments in developing new knowledge and new technological capabilities as well as investments in the institutions and processes that help ensure such advances are actually used to inform policy decisions. There is, in particular, an important role for federal support of basic climate-related research and pre-commercial technology development, because private firms and individuals typically do not stand to benefit commercially from, and thus are unlikely to invest in, such efforts. As discussed below, improved coordination among the federal agencies involved, and between domestic and foreign research efforts, can increase the returns on research spending.

Advancing Scientific Understanding and Technological Development

As discussed in NRC, *Advancing the Science*, scientific research in the United States and around the world has greatly enhanced our understanding of climate change and its causes and effects. The U.S. Global Change Research Program has played an important role in this effort. However, because important questions remain unanswered and because the stakes are so high, it is imperative to continue research, in both the biophysical and social sciences, aimed at increasing our understanding of climate change processes, enhancing our ability to predict climate change and its impacts on critical systems (e.g., agriculture, water resources, urban infrastructure, ecological systems), and increasing our understanding of how people and institutions are affected by and respond to those impacts.

Together with research that enhances fundamental understanding of the climate system, there is a need for research that generates additional options for limiting climate change and adapting to its impacts, and research on how to effectively inform decision making through decision support tools and practices. For example, the development of new technologies for reducing GHG emissions should be accompanied by studies aimed at understanding barriers to their implementation. Adaptation responses can be improved through research on methods for assessing vulnerability and on integrative approaches for responding to the impacts of climate change in interaction with other stresses.

Improving observational and monitoring systems, and developing mechanisms that ensure relevant results are available to key decision makers, can yield a substantial return in the form of better decisions (conversely, in the absence of such information, important decisions would have to be made while "flying blind"). Future decision-support efforts can be improved by research on risk communication and risk management processes, by improved understanding of the factors that facilitate or impede decision making, and by analysis of information needs and existing decision-support activities. Among other things, these efforts will require the development and testing of new analytical approaches and integrative models.

If GHG emissions are to be substantially reduced in a growing U.S. economy, research on climate-friendly energy technologies, and factors affecting their adoption, must be an important element of the nation's R&D portfolio. It often takes decades to bring new energy technologies to market and to deploy them widely, and thus in the near term the U.S. energy system will necessarily rely on technologies that are in (at best) pre-commercial development today. Because the benefits of basic research or pre-commercial technology development in the energy sector cannot be completely captured by the entity that performs it, federal government support of these activities is appropriate, while at the same time recognizing that it is important for government R&D dollars be spent with a willingness to take risks and to focus on long-term benefits.

The federal government plays an essential role in the development of climate-friendly technologies, but government programs are generally ill-suited for translating research results into commercial products. Those final stages of the innovation process are almost always best left to the private sector. But the private sector will not invest significant resources in the design and marketing of climate-friendly products unless it can reasonably expect that there will be demand for them. Because the climate-related benefits of low-carbon technologies are not reflected in market prices, without a substantial and rising carbon price, or some other guarantee of a commercial market, the private sector is likely to under-invest in bringing new low-carbon technologies to market. There is likewise a need, in some instances, for policies to help overcome obstacles to technology commercialization. For example, large-scale demonstration programs may be necessary to reduce uncertainties regarding the cost of new-generation nuclear generating units. Similarly, the large-scale deployment of carbon capture and storage would require the creation of a comprehensive legal framework governing the transportation and underground storage of CO_2.

Research on adaptation is another important component of a balanced R&D portfolio. Just as it is appropriate for the federal government to support research on earthquake-resistant building codes (that can be applied by many local governments, none of which could afford the cost of such research alone), so it is appropriate for the federal government to support research on standards for the design of roads, bridges, and other structures that will perform well under a variety of possible future climates.

Other important adaptation-related research topics include, for instance, improving methods for assessing vulnerability and developing crop varieties and farming methods that can perform well in a range of possible future climate conditions.

RECOMMENDATION 3: The federal government should maintain an integrated, coordinated, and expanded portfolio of research programs with the dual aims of increasing our understanding of the causes and consequences of climate change and enhancing our ability to limit climate change and to adapt to its impacts. In order for federal spending on climate-related R&D to yield optimal returns, it must be effectively coordinated across the many different federal agencies involved, and funding must be relatively stable over time.

Making New Knowledge Pay Off Through Effective Information Systems

An effective national response to climate change demands informed decision making that is based on reliable, understandable, and timely climate-related information tailored to the needs of decision makers in different levels of government, the private sector, educators, and the public. Information systems and services are essential to the effective management of climate risks, because they allow decisions to be modified in response to changing conditions and new information through ongoing evaluation and assessment of policies and actions. Good information systems also underpin effective communication.

NRC, *Informing Effective Decisions* identifies several key aspects of information support to help decision makers develop effective responses to climate change, including the following:

- Information on climate change, vulnerability, and impacts in different regions and sectors is needed for formulating adaptation strategies and understanding how GHG emissions reductions may reduce risks.
- Institutions and mechanisms for "climate services" (i.e., the timely production and delivery of useful climate data, information, and knowledge to decision makers) must respond to the needs of users, be understandable and easily accessible, and be based on the best available science.
- Information for GHG management and accounting—such as establishing emission baselines and supporting monitoring, reporting, and verification—are essential to legitimize market-based systems, as well as voluntary commitments by governments and the private sector.
- Information on energy efficiency and GHG emissions (e.g., conveyed through

product labeling) can help encourage consumer purchasing and behavioral changes.

- Public communication must rest on high quality information that clearly conveys climate science and climate choices and that is seen as coming from trusted sources.
- Because many climate-related choices are occurring in an international context (e.g., in the case of global agricultural and trade systems)—it is essential for the United States to support international information systems and assessments.

In a policy area as complex and rapidly changing as climate change, sound iterative risk management requires institutions with the ability and responsibility to monitor new learning and to make it available in understandable, relevant form to decision makers in the public and private sectors. There is much to be gained by sharing information about what works and does not work, both within the United States and internationally. Existing institutions perform some of these functions, but none provides an ideal model for performing the range of tasks described above, and few effectively engage nonfederal actors.

There are a variety of mechanisms that could be developed for carrying the sort of periodic reporting effort described above. NRC, *Limiting the Magnitude* suggests, as one example, a process in which the President periodically reports to Congress on key developments affecting our nation's response to climate change. This process can be seen as analogous to the Economic Report of the President, prepared annually by the Council of Economic Advisers. It could build upon existing mechanisms for periodic reporting on climate change information (e.g., the annual GHG emissions inventory carried out by the EPA, the U.S. Climate Action report organized by the State Department as input to the UNFCCC, the *Our Changing Planet* report compiled by the USGCRP), and it may include updates on factors such as:

- national and global emissions trends, and their relationship to developments in our understanding of climate change science (including reporting on whether the United States is making sufficient progress toward meeting its GHG budget);
- energy market developments and trajectories;
- the implementation status, costs, and effectiveness of GHG emission-reduction policies;
- the status of the development and deployment of key technologies for reducing GHG emissions;
- the distributional consequences of emission-reduction policies across income groups and regions of the country;

- developments in understanding of climate change impacts and vulnerability to those impacts; and
- updates of adaptation plans and actions underway at federal, state, and local levels.

A wide array of actors in state and local governments, nongovernmental organizations, and the private sector are already playing important roles and should continue to be involved, in the enterprise of collecting and sharing climate-related information (see NRC, *Informing Effective Decisions* for details). But there are a number of areas where federal-level leadership is of particular importance. This includes, for instance, the issuance of federal guidelines for gathering and reporting of key climate-related information, to help ensure the legitimacy and comparability of information being collected by different parties. It also includes monitoring relevant developments internationally and ensuring information access for especially vulnerable populations. Current examples of federal leadership include NOAA and NASA's roles in collecting basic observations of atmospheric, oceanic, and land-surface changes (with regional and private sector actors adding local detail and value-added products); and the EPA's role in collecting and evaluating emission inventory data. Such efforts are clearly valuable in national debates about whether climate change is happening, and whether responses are effective.

RECOMMENDATION 4: The federal government should lead in developing, supporting, and coordinating the information systems needed to inform and evaluate America's climate choices, to ensure legitimacy and access to climate services, greenhouse gas accounting systems, and educational information. To help garner public trust, the design and implementation of any such information systems should be transparent and subject to periodic independent review.

Engaging the Broader Community

As discussed in Chapter 4, establishing processes that bring together scientific / technical experts and government officials with key stakeholders in the private sector and the general public is essential for the success of an iterative risk management approach to addressing climate change. This is because these other stakeholders make important contributions to mitigation and adaptation efforts through their daily choices; because they are an important source of information and perspectives in assessing policies options and in setting priorities for research and development; and because they determine the direction and viability of most governmental policies over the long term.

Processes for engaging stakeholders in learning and deliberation take many forms (for example, Box 4.2 discusses the value of analytic deliberative processes). A substantial research literature, summarized in a number of NRC reports, identifies design principles and tools for implementing such engagement processes. These principles include, for instance, the need to be collaborative and broad based, to combine deliberation with analysis, to ensure transparency of information and analysis, to attend to both facts and values, to explicitly address assumptions and uncertainties, to provide a means of inquiring into official analyses, and to allow iterative reassessment of prior conclusions based on new information.[26] Federal agencies and other organizations that can provide scientific analyses for informing climate choices could do so through direct engagement with the regions, sectors, and constituencies they serve. There are many examples already under way, ranging from national networks such as NOAA's Regional Integrated Science and Assessment Centers to ad hoc community-level dialogues.

RECOMMENDATION 5: The nation's climate change response efforts should include broad-based deliberative processes for assuring public and private-sector engagement with scientific analyses, and with the development, implementation, and periodic review of public policies. Such processes can be initiated by federal agencies, state or local governments, the private sector, or non-profit organizations—linking the organizations that can provide relevant scientific analyses with the constituencies they are best suited to serve, and engaging those who are most affected by a given decision.

INTERNATIONAL ENGAGEMENT

The United States has a strong national interest in ensuring an effective global response to climate change, because if domestic GHG emissions reductions are to be effective in actually limiting climate change, they must be accompanied by significant emission reductions from all major emitting countries. Also, the United States can be deeply affected by climate change impacts occurring elsewhere, given the degree to which different nations are linked by shared natural resources (e.g., fisheries, cross-border river systems), migration of species, diseases vectors, and human populations, and linked economic and trade systems. The United States can magnify the returns on its climate-related investments by a thoughtful strategy of international engagement that encompasses all the various activities discussed in earlier sections of this chapter.

In the committee's judgment, serious U.S. emission-reduction efforts and effective participation in international negotiations are necessary conditions for stimulating substantial global emission reductions. The difficult and unwieldy nature of the UNFCCC process underlies the need for U.S. diplomatic efforts to be enhanced by continued involvement in the Major Emitters Forum and other bilateral and multilateral settings. As discussed earlier, the use of international offsets, in which a U.S. firm pays for and receives credit for relatively inexpensive emissions reductions (relative to some baseline) in another country, may play a constructive role in engaging developing nations; but only experience will tell if these approaches will be sufficient to induce substantial global emissions reductions.

International agreements to limit GHG emissions (and to enhance GHG sinks through land-use practices) will require rigorous methods to accurately estimate these emissions, monitor their changes over time, and verify them with independent data. To help assure that such efforts are carried out with transparency, consistency, and proper quality assurance in all countries, there is a need for active U.S. participation in international cooperative efforts, including financial and technical assistance for developing countries that lack the needed resources and expertise. In 2010, the NRC assessed existing capabilities for estimating and verifying GHG emissions and identified ways to improve these capabilities through strategic near-term investments.[27]

The United States also has much to gain from actively participating in international adaptation efforts, particularly those involving developing nations. It is in the nation's interest to limit the potentially destabilizing impacts of climate change in the developing world, and it can be argued that our large contribution to current and historic global GHG emissions gives us some responsibility to assist those whose vulnerabilities exceed their resources. Moreover, active U.S. participation in international adaptation programs could enable us to learn from the effective programs of others. Some efforts could be collaborative, such as research on drought-resistant agriculture for tropical regions. Some efforts might be carried out through existing treaties and development assistance programs, for instance, various UN and World Bank programs, as well as existing UNFCCC adaptation funding programs. Additional mechanisms may be needed, however, for multilateral exploration of techniques and technologies that support adaptation, as well as for communication and trust-building. Durable institutional arrangements will be necessary for long-term success.

The United States also has much to gain from international engagement in scientific research and technology development. Understanding and responding to the risks of climate change requires ongoing efforts to collect and evaluate a vast array of information from around the world. In addition to observations of the climate itself, this includes data on relevant socioeconomic indicators, on emissions monitoring and verification activities, and on best practices in climate change limiting and adaptation efforts. Some relevant information can be gathered by top-down physical observations (e.g., satellite remote sensing), but some will require the bottom-up collection and synthesis of detailed local-scale monitoring. Almost all such efforts are beyond the means of any single country and can only be effectively advanced through international cooperation, in which U.S. leadership could well prove critical.

Carefully targeted and coordinated R&D efforts can help enable developing nations to achieve both the economic growth necessary to alleviate poverty and the GHG emissions reductions necessary to limit future climate change. If today's poor nations rely only on currently available technologies in their drive to approach the living standards of today's rich nations, dramatic increases in global CO_2 emissions are inevitable. To reduce global emissions, developing nations must travel a less carbon-intensive development path; and it is in U.S. interest to help facilitate their efforts to follow these alternative paths. In addition, U.S. firms could gain valuable technology and market access from participation in international efforts to develop low-carbon technologies.

RECOMMENDATION 6: The United States should actively engage in international-level climate change response efforts: to reduce greenhouse gas emissions through cooperative technology development and sharing of expertise, to enhance adaptive capabilities (particularly among developing nations that lack the needed resources), and to advance the research and observations necessary to better understand the causes and effects of climate change.

TOWARD AN INTEGRATED NATIONAL RESPONSE

The different types of actions presented in this chapter as part of America's climate choices are not separate and distinct. Rather, actions in one category can directly affect actions in others—for better or for worse. For example, the more successful efforts are to reduce GHG emissions, the less climate change there will be to adapt to, and the more time will be available to adjust. The more we advance basic understanding of the climate system, the more effective our responses will be. Table 5.1 illustrates these and other linkages among the different elements of a comprehensive response strategy. A comprehensive national response strategy that effectively integrates these different elements, however, presents significant challenges of coordinating across different levels of government, across different types of organizations (within and outside government), and across different types of response functions. Each of these challenges is discussed below.

TABLE 5.1 Matrix of Interdependencies Among the Different Elements of a National Response to Climate Change

Will strengthen this element because…

	Limiting	Adapting	Advancing Science & Technology	Informing
Limiting		There may be less stringent, disruptive requirements (and thus lower costs) for adapting to climate change impacts.	There may be less pressure to develop risky and/or expensive technologies for coping with impacts.	The decision environment may be less contentious if the severity of climate change can be limited.
Adapting	Any given degree of climate change may be associated with less severe impacts and disruptions of human and natural systems.		There may be less pressure to develop risky and/or expensive technologies for limiting climate change (e.g., some forms of geoengineering).	The decision environment may be less contentious if communities and key sectors are prepared to deal with impacts.
Advancing Science & Technology	R&D could help identify more and better options for limiting climate change.	R&D could help provide more adaptation options and more knowledge about their implications.		The knowledge base for informing decisions may be more complete, and the knowledge base about how to most effectively inform may allow better information flow.
Informing	Effective options for limiting climate change may be more widely deployed and used.	Effective options for adapting to climate change may be more widely deployed and used.	Science may be more attuned to decision needs, and public support for advances in science is likely to increase.	

Advancing this element…..

75

Coordination Across Levels of Government

Efforts to limit and adapt to climate change present different types of coordination challenges. Adaptation choices will be made largely by state and local governments and the private sector. The federal government can play an important leadership role by providing widely useful knowledge and information, but an effective *national* adaptation strategy will be based not on top-down federal directives. Rather, it will be based on coordination and information sharing across levels of government and between public and private sectors.

The federal role is more obviously and critically important in limiting GHG emissions. The relevant domestic costs and benefits can be fully aggregated only at the national level, and strong federal action will be necessary to achieve large U.S. emission reductions and to sustain an effective, balanced R&D program. Nevertheless, many states and localities have taken significant early steps to limit emissions, and some important limiting options, such as revising building codes, changing land-use patterns, and reconfiguring transportation systems, are within the traditional authority and expertise of state and local governments.

Effective coordination requires carefully balancing federal with state and local authority and promoting regulatory flexibility across jurisdictional boundaries where it is sensible to do. This includes, for instance, allowing states the option of regulating GHG emissions more stringently than federal law (in which case, the state is shifting more of the burden of meeting national goals onto its own residents). There is generally little to be gained by preempting such state regulations, as long as one can avoid standard-setting that fragments the national market among numerous states with differing regulations. Perhaps most importantly, efforts of state and local governments to reduce GHG emissions or adapt to climate change provide valuable policy experiments from which decision makers at other levels can draw useful lessons.

Coordination can also take the form of providing federal incentives (or removing disincentives) for action by states and localities. This includes, for instance: ensuring that states and localities have sufficient resources to implement and enforce significant new regulatory burdens placed on them by federal policy makers (e.g., national building standards); ensuring that new federal directives do not disadvantage states and localities that have taken early action to reduce emissions; and providing incentives for adaptation planning across jurisdictions and sectors.

Coordination Across Organizations

Dozens of federal agencies and other organizations are carrying out research, making decisions, and taking action on climate change through a host of existing programs and authorities. These include, for example, adaptation on federal lands (Departments of Agriculture, Interior, and Defense); research on climate change and related impacts (many agencies); research and development for technologies to respond to climate change (Department of Energy); information provision (Energy Information Administration, National Oceanic and Atmospheric Administration, Environmental Protection Agency); and regulation of automobile efficiency and GHG emissions (Department of Transportation, Environmental Protection Agency). Many additional organizations will likely engage as national strategies for limiting and adapting to climate change emerge.

Although the various activities carried out through these different programs are inextricably linked, they are managed largely as separate, isolated activities across the federal government. For example, many departments and agencies that are or will be engaged in climate response (e.g., Federal Emergency Management Agency, Department of Housing and Urban Development, Department of Energy) have not been part of the U.S. Global Change Research Program (USGCRP) and lack sufficient communication with the federal agencies that are developing knowledge they need. The USGCRP and the Climate Change Adaptation Task Force have largely been confined to convening representatives of relevant agencies and programs for dialogue, without mechanisms for making or enforcing important decisions and priorities. Moreover, even the USGCRP and the Climate Change Technology Program together do not appear sufficient for effectively coordinating the full portfolio of research needed to support climate change response efforts.

One can look to other major policy arenas (e.g., public health, national security) and to other countries for examples of different coordinating mechanisms that have been employed with varying degrees of success. Some common models include:

- Giving one federal agency full responsibility and authority to lead and co-ordinate activities across the federal government (e.g., as has occurred with climate change in the United Kingdom and Australia);
- Creating a White House staff position tasked with directly advising the President on policy decisions and leading coordination efforts (e.g., a climate "czar");
- Establishing a new executive branch organization, staffed by senior-level officials from other relevant government bodies, responsible for coordinating policy and advising the President (e.g., the National Security Council model).

A detailed evaluation of the pros and cons associated with each of these different organizational models is beyond the scope of this report, but the next section discusses the general capabilities and responsibilities that would be most important for any such coordinating entity.

Coordination Across Functions

Many previous NRC studies have offered guidance on how to ensure that decision makers are informed by the best relevant scientific and technical analysis;[28] but in the context of climate change, efforts to actually do so are in their infancy. Traditionally, climate change research efforts have been organized predominantly around priorities defined by advancing scientific understanding, which do not necessarily match the needs of affected decision makers. Meeting the coordination challenge of linking knowledge to action will require sustained efforts from decision makers at all levels, but there is a particularly strong need for federal leadership. Federal agencies can create organizations to perform coordination functions for particular regions or sectors, as NOAA has done with its Regional Integrated Sciences and Assessments Program, and DOI is planning to do with its Landscape Conservation Councils and Climate Science Centers. They can also support networks that link decision makers within a region or sector to each other and to decision-relevant knowledge, can facilitate processes to collect and analyze data on climate response efforts around the country, and can communicate the lessons from objective assessments of these efforts, thus enabling decision makers to learn from each other's experiences.

In summary, the following are some essential coordination challenges that a national climate change response effort will need to address:

- Ensuring that federal actions facilitate (or at a minimum, do not impede) effective nonfederal actions for mitigation and adaptation;
- Developing a clear division of labor among federal agencies and a process to monitor how well this division of labor is functioning over time;
- Ensuring decision support for constituencies that do not have a particular government agency or program responsible for providing such information; and
- Linking science, decision support, and resource management functions within the federal response to climate change.

To address such wide-ranging challenges, any institution with major responsibility for coordinating our nation's climate change response efforts will need to have several key features, including: authority to set priorities and to turn these priorities into resource allocation decisions; sufficient budgetary resources to actually implement allocation decisions; personnel who both understand climate science and understand the needs of climate-affected decision makers; mechanisms for monitoring the organization's performance, in order to improve over time; and processes to ensure accountability to the parties that use information developed or shared by the institution.

RECOMMENDATION 7: The federal government should facilitate coordination of the many interrelated components of America's response to climate change with a process that identifies the most critical coordination issues and recommends concrete steps for how to address these issues. Coordination and possible reorganization among federal agencies will require attention from the highest levels of the executive branch and from Congress. In areas of mixed federal and non-federal responsibility, the federal government's leadership role should emphasize support and facilitation of decentralized responses at lower levels of government and in the private sector.

CHAPTER CONCLUSION

Responding to the risks of climate change is one of the most important challenges facing the United States today. Unfortunately, there is no "magic bullet" for dealing with this issue; no single solution or set of actions that can eliminate the risks we face. America's climate choices will involve political and value judgments by decision makers at all levels. These choices, however, must be informed by sound scientific analyses. This report recommends a diversified portfolio of actions, combined with a concerted effort to learn from experience as those actions proceed, to lay the foundation for sound decision-making today and expand the options available to decision makers in the future. Doing so will require political will and resolve, innovation and perseverance, and collaboration across a wide range of actors.

Notes and References

SUMMARY

1. The analyses in this report focus primarily on energy-related carbon dioxide (CO_2) emissions, which constitute roughly 83 percent of total U.S. greenhouse gas emissions (U.S. Environmental Protection Agency [EPA], *Draft Inventory of U.S. Greenhouse Gas Emissions and Sinks: 1990-2009*. EPA 430-R-11-005 [Washington, D.C.: EPA, 2011, available at *http://www.epa.gov/climatechange/emissions/usinventoryreport.html*, accessed March 2, 2011]). As discussed in Chapter 1, however, there are other long-lived greenhouse gases, as well as shorter-lived gas and particulate compounds, that contribute to climate change and offer mitigation opportunities.

CHAPTER 1

1. United Nations Framework Convention on Climate Change (UNFCCC), i.e., the "Rio declaration" ratified by the United States in 1992.
2. EPA, *Draft Inventory*.
4. J. M. Broder, "Emissions fell in 2009, showing impact of recession," (*New York Times*, Feb 16, 2011, available at *http://www.nytimes.com/2011/02/17/science/earth/17emit.html?_r=1*, accessed April 11, 2011).
5. Energy Information Administration (EIA), *International Energy Outlook,* Report #:DOE/EIA-0484(2010) (Washington, D.C.: U.S. Department of Energy, 2010, available at *http://www.eia.doe.gov/oiaf/ieo/*, accessed March 4, 2011).
6. S. J. Davis, K. Calderia, and D. Matthews, "Future CO_2 emissions and climate change from existing energy infrastructure" (*Science* 10 329[5997]:1330-1333, 2010, doi: 10.1126/science.1188566).
7. International Energy Agency. 2010. *World Energy Outlook 2010* (Paris, France: International Energy Agency, 2010).
8. *http://unfccc.int/resource/docs/2009/cop15/eng/l07.pdf.*
9. Massachusetts v. Environmental Protection Agency, 549 U.S. 497. 2007 (*http://www.supremecourt.gov/opinions/06pdf/05-1120.pdf*).
10. *http://www.epa.gov/NSR/actions.html#may10.*
11. See: A. J. Hoffman, "Climate change strategy: The business logic behind voluntary greenhouse gas reductions" (*California Management Review* 47[3]:21-46, 2005); NRC, *America's Climate Choices: Informing Effective Decisions on Climate Change* (Washington, D.C.: National Academies Press, 2010), Table 2.3; and reports of the World Business Council on Sustainable Development (http://www.wbcsd.org/), the U.S. Climate Action Partnership (http://www.us-cap.org/), the American Energy Innovation Council (*http://www.americanenergyinnovation.org/*), and the Business Environmental Leadership Council (*http://www.pewclimate.org/companies_leading_the_way_belc*).
12. *http://www.usmayors.org/climateprotection/revised/.*
13. A renewable portfolio standard (RPS) mandates the use of a given percentage of renewable energy sources in an overall energy mix for example in the electricity sector.
14. *http://www.pewclimate.org/what_s_being_done/in_the_states/rps.cfm*; and for a comprehensive map/listing of relevant state efforts, see: http://www.pewclimate.org/states-regions.
15. *http://www.theclimateregistry.org/, http://www.climateregistry.org/, https://www.cdproject.net/.*
16. For example, King County, Washington, worked with the Climate Impacts Group at the University of Washington and ICLEI: Local Governments for Sustainability to produce a handbook in 2007, *Preparing for Climate Change: A Guidebook for Local, Regional, and State Governments.*

17.   Department of Defense, Quadrennial Defense Review Report (Washington, D.C.: U.S. Department of Defense, 2010, available at *http://www.defense.gov/qdr/*, accessed March 1, 2011); D. C. Blair, *Annual Threat Assessment of the U.S. Intelligence Community*. Statement for the Record to the Senate Select Committee on Intelligence (2010, available at http://www.dni.gov/testimonies/20100202_testimony.pdf, accessed February 28, 2011); NRC, *Advancing the Science of Climate Change* (Washington, D.C.: National Academies Press, 2010), Chapter 15.

18.   Center for Climate Strategies, *Impacts of Comprehensive Climate and Energy Policy Options on the U.S. Economy* (Washington, D.C.: Johns Hopkins University, 2010, available at *http://advanced.jhu.edu/academic/government/energy-policy-report/*, accessed March 1, 2011); N. M . Bianco and F. T. Litz, *Reducing Greenhouse Gas Emissions in the United States: Using Existing Federal Authorities and State Action* (Washington, D.C.: World Resources Institute, 2010, available at *http://www.wri.org/publication/reducing-ghg-emissions-using-existing-federal-authorities-and-state-action*, accessed March 8, 2011).

19.   M. Betsill and H. Bulkeley, "Cities and the multilevel governance of global climate change" (*Global Governance* 12 [2]:141-159, 2006).

20.   See, for example, E. Ostrom, "A multi-scale approach to coping with climate change and other collective action problems" (*Solutions* 1:27-36, 2010).

CHAPTER 2

1.   For a historical overview of the science of climate change, see S. R. Weart, *The Discovery of Global Warming* (Cambridge, MA: Harvard University Press, 2008).

2.   According to IPCC (*Climate Change 2007: The Physical Science Basis. Contribution of Working Group I to the Fourth Assessment Report of the Intergovernmental Panel on Climate Change*, eds. S. Solomon, D. Qin, M. Manning, Z. Chen, M. Marquis, K. B. Averyt, M. Tignor, and H. L. Miller [Cambridge, UK: Cambridge University Press, 2007]): "Warming of the climate system is unequivocal, as is now evident from observations of increases in global average air and ocean temperatures, widespread melting of snow and ice, and rising global average sea level."

3.   The temperature of the lower atmosphere is measured directly by instruments mounted on weather balloons (see, e.g., M. P. McCarthy, H. A. Titchner, P. W. Thorne, S. F. B. Tett, L. Haimberger, and D. E. Parker, "Assessing bias and uncertainty in the HadAT-adjusted radiosonde climate record" [*Journal of Climate* 21(4):817-832, 2008]) and indirectly by satellites that measure the energy radiated upward from the Earth at specific wavelengths (J. R. Christy, R. W. Spencer, W. D. Braswell, "MSU tropospheric temperatures: Dataset construction and radiosonde comparisons" [*Journal of Atmospheric and Oceanic Technology* 17:1153-1170, 2000]; J. R. Christy, R. W. Spencer, W. B. Norris, and W. D. Braswell, "Error estimates of version 5.0 of MSUAMSU bulk atmospheric temperatures" [*Journal of Atmospheric and Oceanic Technology* 20:613-629, 2003]; C. A. Mears and F. J. Wentz, "Construction of the remote sensing systems V3.2 atmospheric temperature records from the MSU and AMSU microwave sounders" [*Journal of Atmospheric and Oceanic Technology* 26:1040-1056, 2009, doi: 10.1175/2008JTECHA1176.1]). Trends in these data have been assessed in detail by the U. S. Climate Change Science Program (CCSP (*Temperature Trends in the Lower Atmosphere: Steps for Understanding and Reconciling Differences*, Synthesis and Assessment Product 1.1, eds. T. R. Karl, S. J. Hassol, C. D. Miller, and W. L. Murray [Washington, D.C.: National Oceanic and Atmospheric Administration, 2006]) and the IPCC (*Climate Change 2007 WG1*, Summary for Policymakers).

4.   The warming of the uppermost 700 meters of the world oceans is documented in S. J. Levitus, I. Antonov, T. P. Boyer, R. A. Locarnini, H. E. Garcia, and A. V. Mishonov, "Global ocean heat content 1955-2008 in light of recently revealed instrumentation problems" (*Geophysical Research Letters* 36:L07608, 2009).

5.   According to the IPCC (*Climate Change 2007*, Ch9): "anthropogenic forcing has likely contributed to recent decreases in Arctic sea ice extent and to glacier retreat."

6.   IPCC (*Climate Change 2007 WG1*, Ch9) concludes "It is likely that anthropogenic forcing has contributed to the general warming observed in the upper several hundred meters of the ocean during the latter half of the 20th century."

7.	According to IPCC (*Climate Change 2007: Impacts, Adaptation and Vulnerability. Contribution of Working Group II to the Fourth Assessment Report of the Intergovernmental Panel on Climate Change*, eds. M. L. Parry, O. F. Canziani, J. P. Palutikof, P. J. van der Linden, and C. E. Hanson [Cambridge, UK: Cambridge University Press, 2007]): "There is very high confidence [about 8 out of 10 chance of being correct] … that recent warming is strongly affecting terrestrial biological systems" and "high confidence, based on substantial new evidence, that observed changes in marine and freshwater biological systems are associated with rising water temperatures, as well as related changes in ice cover, salinity, oxygen levels and circulation."

8.	According to IPCC (*Climate Change 2007 WG1*, Summary for Policymakers): "Most of the observed increase in global average temperatures since the mid-20th century is very likely [greater than 90 percent likelihood] due to the observed increase in anthropogenic greenhouse gas concentrations."

9.	D. Lüthi, M. Le Floch, B. Bereiter, T. Blunier, J.-M. Barnola, U. Siegenthaler, D. Raynaud, J. Jouzel, H. Fischer, K. Kawamura, and T. F. Stocker, "High-resolution carbon dioxide concentration record 650,000-800,000 years before present" (*Nature* 453[7193]:379-382, 2008, doi:10.1038/nature06949).

10.	IPCC (*Climate Change 2007 WG1*, Summary for Policymakers): "The primary source of the increased atmospheric concentration of carbon dioxide since the pre-industrial period results from fossil fuel use, with land-use change providing another significant but smaller contribution." This statement is based in part on the fossil fuel emissions data in Figure 1.2; in part on estimates of the other sources as "sinks" of atmospheric carbon dioxide like those provided by the Global Carbon Project (Le Quéré, C. M. R. Raupach, J. G. Canadell, G. Marland, L. Bopp, P. Ciais, T. J. Conway, S. C. Doney, R. A. Feely, P. Foster, P. Friedlingstein, K. Gurney, R. A. Houghton, J. I. House, C. Huntingford, P. E. Levy, M. R. Lomas, J. Majkut, N. Metzl, J. P. Ometto, G. P. Peters, I. C. Prentice, J. T. Randerson, S. W. Running, J. L. Sarmiento, U. Schuster, S. Sitch, T. Takahashi, N. Viovy, G. R. van der Werf, and F. I. Woodward, "Trends in the sources and sinks of carbon dioxide" [*Nature Geoscience* 2, 2009, doi: 10.1038/ngeo689]), which indicate that deforestation and other land use changes currently contribute about 12% of total human-induced CO_2 emissions; and in part on the chemical "fingerprints" of CO_2 and other gases in the atmosphere, which can only be explained by the burning of coal, oil, and natural gas (R. F. Keeling, S. C. Piper, A. F. Bollenbacher and J. S. Walker, "Atmospheric CO_2 records from sites in the SIO air sampling network," in *Trends: A Compendium of Data on Global Change* (Oak Ridge, TN: Carbon Dioxide Information Analysis Center, Oak Ridge National Laboratory, U.S. Department of Energy, 2009).

11.	Methane comes from fossil fuel and biomass burning, natural gas management, animal husbandry, rice cultivation, and waste management (S. Houweling, T. Rockmann, I. Aben, F. Keppler, M. Krol, J. F. Meirink, E. J. Dlugokencky, and C. Frankenberg, "Atmospheric constraints on global emissions of methane from plants" [*Geophysical Research Letters* 33:L15821, 2006, doi:10.1029/2006GL026162]). The atmospheric concentration of methane rose sharply through the late 1970s before leveling off at about two-and-a-half times its estimated pre-industrial concentration. Methane levels have risen slightly in each of the past few years (E. J. Dlugokencky, L. Bruhwiler, J. W. C. White, L. K. Emmons, P. C. Novelli, S. A. Montzka, K. A. Masarie, P. M. Lang, A. M. Crotwell, J. B. Miller, and L. V. Gatti, "Observational constraints on recent increases in the atmospheric CH4 burden" [*Geophysical Research Letters* 36:L18803, 2009]) but the reasons for the changes are not completely clear. Nitrous oxide concentrations are steadily increasing primarily as a result of agricultural activities (especially the application of chemical fertilizers), but also a byproduct of fossil fuel combustion and certain industrial process. Halogenated gases include chlorofluorocarbons (CFCs), hydrofluorocarbons, perfluorocarbons, and sulfur hexafluoride, all of which are produced primarily by industrial processes. Many of these compounds also contribute to the depletion of ozone in the stratosphere, which is a related but largely separate environmental problem from climate change (see, e.g., CCSP, *Trends in Emissions of Ozone-Depleting Substances, Ozone Layer Recovery, and Implications for Ultraviolet Radiation Exposure*, Synthesis and Assessment Product 2.4 by the U.S. Climate Change Science Program and the Subcommittee on Global Change Research, eds. A. R. Ravishankara, M. J. Kurylo, and C. A. Ennis [Asheville, NC: National Oceanic and Atmospheric Administration, 2008]). Water vapor is also an important greenhouse gas, but its concentration in the lower atmosphere is controlled by the rates of evaporation and precipitation, which are natural processes that are much more strongly influenced by changes in atmospheric temperature and circulation than by human activities directly.

12. T. P. Barnett, D. W. Pierce, K. M. AchutaRao, P. J. Gleckler, B. D. Santer, J. M. Gregory, and W. M. Washington ("Penetration of human-induced warming into the world's oceans" [*Science* 309(5732):284-287, 2005, doi: 10.1126/science.1112418]) conclude that the observed warming trend cannot be explained by the release of heat stored in the deep ocean or other climate system components, while IPCC (*Climate Change 2007 WG1*) concludes that "it is extremely unlikely (<5%) that the global pattern of warming during the past half century can be explained without external forcing, and very unlikely (<10%) that it is due to known natural external causes alone." The latter conclusion is based in part on the fact that models of the climate system are able to reproduce the observed spatial and temporal pattern of warming when human-induced GHG and aerosol emissions are included in the simulation, but not when only natural climate forcing factors are included (IPCC, *Climate Change 2007 WG1*, Ch8). Reconstructions of solar activity based on historical records and other sources suggest that the amount of energy reaching Earth from the sun may have increased slightly during the late 19th and early 20th century, possibly contributing to some of the warming observed in the first few decades of the 20th century, but satellite observations show definitively that solar output has not increased overall during the last 30 years (J. L. Lean and T. N. Woods, "Solar total and spectral irradiance: Measurements and models," in *Heliophysics: Evolving Solar Physics and the Climates of Earth and Space*, eds. C. J. Schrijver and G. Siscoe (Cambridge, UK: Cambridge University Press, 2010).

13. According to USGCRP (*Global Climate Change Impacts in the United States,* eds. T. R. Karl, J. M. Melillo, and T. C. Peterson [Cambridge, U.K.: Cambridge University Press, 2009]): ". . . scientists have established causal links between human activities and the changes in snowpack, maximum and minimum temperature, and the seasonal timing of runoff over mountainous regions of the western United States" (see also T. P. Barnett, D. W. Pierce, H. G. Hidalgo, C. Bonfils, B. D. Santer, T. Das, G. Bala, A. W. Wood, T. Nozawa, A. A. Mirin, D. R. Cayan, and M. D. Dettinger, "Human-induced changes in the hydrology of the western United States" [*Science* 319(5866):1080-1083, 2008]; D. W. Pierce, T. P. Barnett, H. G. Hidalgo, T. Das, C. Bonfils, B. D. Santer, G. Bala, M. D. Dettinger, D. R. Cayan, A. Mirin, A. W. Wood, and T. Nozawa, "Attribution of declining western U.S. snowpack to human effects" [*Journal of Climate* 21(23):6425-6444, 2008]; and C. Bonfils, B. D. Santer, D. W. Pierce, H. G. Hidalgo, G. Bala, T. Das, T. P. Barnett, D. R. Cayan, C. Doutriaux, A. W. Wood, A. Mirin, and T. Nozawa, "Detection and attribution of temperature changes in the mountainous western United States" [*Journal of Climate* 21(23):6404-6424, 2008]).

14. According to IPCC (*Climate Change 2007 WG1*, Summary for Policymakers): "At continental, regional and ocean basin scales, numerous long-term changes in climate have been observed. These include changes in Arctic temperatures and ice, widespread changes in precipitation amounts, ocean salinity, wind patterns, and aspects of extreme weather including droughts, heavy precipitation, heat waves, and the intensity of tropical cyclones."

15. E.g., J. Oerlemans, "Extracting a climate signal from 169 glacier records" (*Science* 308[5722]:675-677, 2005); R. Thomas, E. Frederick, W. Krabill, S. Manizade, and C. Martin, "Progressive increase in ice loss from Greenland" (*Geophysical Research Letters* 33[10], 2006); E. Rignot, J. E. Box, E. Burgess, and E. Hanna, "Mass balance of the Greenland ice sheet from 1958 to 2007" (*Geophysical Research Letters* 35[20], 2008); I. Velicogna, "Increasing rates of ice mass loss from the Greenland and Antarctic ice sheets revealed by GRACE" (*Geophysical Research Letters* 36, 2009).

16. IPCC (Climate Change 2007 WG1, Ch5), based on data from, J. A. Church and N. J. White, "A 20th century acceleration in global sea-level rise" (*Geophysical Research Letters* 33[1], 2006); S. J. Holgate and P. L. Woodworth, "Evidence for enhanced coastal sea level rise during the 1990s" (*Geophysical Research Letters* 31[7]:L07305, 2004); and E. W. Leuliette, R. S. Nerem, and G. T. Mitchum, "Calibration of TOPEX/Poseidon and Jason altimeter data to construct a continuous record of mean sea level change" (*Marine Geodesy* 27[1-2]:79-94, 2004).

17. IPCC (*Climate Change 2007 WG1*, Ch4) update of R. D. Brown, "Northern Hemisphere snow cover variability and change, 1915-97" (*Journal of Climate* 13:2339-2355, 2000).

18. S. K. Min, X. Zhang, F. W. Zwiers, and T. Agnew ("Human influence on Arctic sea ice detectable from early 1990s onwards" [*Geophysical Research Letters* 35(21), 2008]) conclude that the overall decline in Arctic sea ice since 1979 is very likely due to human-induced warming; discussion of recent sea ice trends (which have been attributed to changes in wind patterns as well as to warming) can also be found in J. E. Overland, M. Wang, and S. Salo, "The recent Arctic warm period" (*Tellus Series A—Dynamic Meteorology and Oceanography* 60[4]:589-597, 2008); X. D. Zhang, A. Sorteberg, J. Zhang, R. Gerdes, and J. C. Comiso, "Recent radical shifts of atmospheric circulations and rapid changes in Arctic climate system" (*Geophysical Research Letters* 35[22], 2008); J. Stroeve, M. M. Holland,

W. Meier, T. Scambos, and M. Serreze, "Arctic sea ice decline: faster than forecast" (*Geophysical Research Letters* 34[9]:L09501, 2007, doi:10.1029/2007GL029703); and M. C., Serreze, M. M. Holland, and J. Stroeve, "Perspectives on the Arctic's shrinking sea-ice cover" (*Science* 315[5818]:1533-1536, 2007, doi:10.1126/science.1139426). In the Southern Hemisphere, sea ice cover has actually increased slightly over the past several decades; this trend has been attributed to changes in atmospheric circulation associated with stratospheric ozone depletion (D. J. Cavalieri and C. L. Parkinson, "Antarctic sea ice variability and trends, 1979-2006" [*Journal of Geophysical Research-Oceans* 113(C7):C07004, 2008]; J. Turner, J. C. Comiso, G. J. Marshall, T. A. Lachlan-Cope, T. Bracegirdle, T. Maksym, M. P. Meredith, Z. M. Wang, and A. Orr, "Non-annular atmospheric circulation change induced by stratospheric ozone depletion and its role in the recent increase of Antarctic sea ice extent" [*Geophysical Research Letters* 36:L08502, 2009, doi: 10.1029/2009GL037524]).

19. According to IPCC (*Climate Change 2007 WG2*, Summary for Policymakers): "Observational evidence from all continents and most oceans shows that many natural systems are being affected by regional climate changes, particularly temperature increases" and "a global assessment of data since 1970 has shown it is likely (>66% likelihood) that anthropogenic warming has had a discernible influence on many physical and biological systems."

20. According to IPCC (*Climate Change 2007 WG1*, Ch4): "The maximum extent of seasonally frozen ground has decreased by about 7% in the Northern Hemisphere from 1901 to 2002, with a decrease in spring of up to 15%." There are substantial regional differences in permafrost warming/melting; permafrost trends in Russia, for example, are documented by H. J. Akerman and M. Johansson, "Thawing permafrost and thicker active layers in sub-arctic Sweden" (*Permafrost and Periglacial Processes* 19[3]:279-292, 2008); V. E. Romanovsky, T. S. Sazonova, V. T. Balobaev, N. I. Shender, and D. O. Sergueev, "Past and recent changes in air and permafrost temperatures in eastern Siberia" (*Global and Planetary Change* 56[3-4]:399-413, 2007), and T. Osterkamp, "Characteristics of the recent warming of permafrost in Alaska" (*Journal of Geophysical Research* 112:F02S02, 2007, doi:10.1029/2006JF000578).

21. The IPCC (*Climate Change 2007 WG1*) report that averaged over all available data, the freezing period of lake and river ice has decreased by 12 days over the past 150 years.

22. As discussed in USGCRP (*Global Climate Change Impacts*, page 85): "As the carbon dioxide concentration in the air increases, more carbon dioxide is absorbed into the world's oceans, leading to their acidification. This makes less calcium carbonate available for corals and other sea life to build their skeletons and shells. If carbon dioxide concentrations continue to rise and the resulting acidification proceeds, eventually, corals and other ocean life that rely on calcium carbonate will not be able to build these skeletons and shells at all. The implications of such extreme changes in ocean ecosystems are not clear, but there is now evidence that … acidification is already occurring. See also NRC, *Ocean Acidification: A National Strategy to Meet the Challenges of a Changing Ocean* (Washington, D.C.: National Academies Press, 2010); and NRC, *Advancing the Science*.

23. USGCRP *Global Climate Change Impacts*.

24. USGCRP (*Global Climate Change Impacts*, pages 28 and 30), based on U.S. Historical Climate Network data from NOAA/NCDC (M. J. Menne and C. N. Williams, Jr., "Homogenization of temperature series via pairwise comparisons" (*Journal of Climate* 22:1700-1717, 2009), http://www.ncdc.noaa.gov/oa/climate/research/ushcn/).

25. USGCRP (*Global Climate Change Impacts*, pages 37 and 149); See also CCSP, *Coastal Sensitivity to Sea-Level Rise: A Focus on the Mid-Atlantic Region, Synthesis and Assessment Product 4.1 by the U.S. Climate Change Science Program and the Subcommittee on Global Change Research*, ed. J. G. Titus, coordinating lead author; E. K. Anderson, D. R. Cahoon, S. Gill, R. E. Thieler, and J. S. Williams, lead authors (Washington, D.C.: U.S. Environmental Protection Agency, 2009) and CCSP, *The Effects of Climate Change on Agriculture, Land Resources, Water Resources, and Biodiversity in the United States, Synthesis and Assessment Product 4.3 by the U.S. Climate Change Science Program and the Subcommittee on Global Change Research*, eds. P. Backlund, A. Janetos, D. Schimel, J. Hatfield, K. Boote, P. Fay, L. Hahn, C. Izaurralde, B. A. Kimball, T. Mader, J. Morgan, D. Ort, W. Polley, A. Thomson, D. Wolfe, M. G. Ryan, S. R. Archer, R. Birdsey, C. Dahm, L. Heath, J. Hicke, D. Hollinger, T. Huxman, G. Okin, R. Oren, J. Randerson, W. Schlesinger, D. Lettenmaier, D. Major, L. Poff, S. Running, L. Hansen, D. Inouye, B. P. Kelly, L. Meyerson, B. Peterson, and R. Shaw (Washington, D.C.: National Oceanic and Atmospheric Administration, 2008).

26. USGCRP (*Global Climate Change Impacts*, page 141); see also CCSP, *SAP 4.3*; Osterkamp, *Characteristics*; and A. Instanes, O. Anisimov, L. Brigham, D. Goering, L. N. Khrustalev, B. Ladanyi, and J. O. Larsen, "Infrastructure: buildings,

support systems, and industrial facilities,"pp. 907-944 in *Arctic Climate Impact Assessment* (Cambridge, UK: Cambridge University Press, 2005).

27. USGCRP (*Global Climate Change Impacts*, page 45); see also B. C. Bates, Z. W. Kundzewicz, S. Wu, and J. P. Palutikof, eds., Climate Change and Water, Technical paper of the Intergovernmental Panel on Climate Change (Geneva, Switzerland: IPCC Secretariat, 2008); P. Mote, A. Hamlet, and E. Salathé, "Has spring snowpack declined in the Washington Cascades?" (*Hydrology and Earth System Sciences* 12[1]:193-206.2008); S. Feng, and Q. Hu, "Changes in winter snowfall/precipitation ratio in the contiguous United States" (*Journal of Geophysical Research* 112:D15109, 2007, doi:10.1029/2007JD008397); I. T. Stewart, D. R. Cayan, and M. D. Dettinger, "Changes toward earlier streamflow timing across western North America" (*Journal of Climate*,18:1136-1155, 2005).

28. USGCRP (*Global Climate Change Impacts*, page 32) reports that "The amount of rain falling in the heaviest downpours has increased approximately 20 percent on average in the past century" in the United States, and that this increase accounts for most of the overall precipitation trend. K. E. Kunkel, P. D. Bromirski, H. E. Brooks, T. Cavazos, A. V. Douglas, D. R. Easterling, K. A. Emanuel, P. Ya. Groisman, G. J. Holland, T. R. Knutson, J. P. Kossin, P. D. Komar, D. H. Levinson, and R. L. Smith "Observed changes in weather and climate extremes," in *Weather and Climate Extremes in a Changing Climate. Regions of Focus: North America, Hawaii, Caribbean, and U.S. Pacific Islands*, eds. T. R. Karl, G. A. Meehl, C. D. Miller, S. J. Hassol, A. M. Waple, and W. L. Murray (Washington, D.C.: U.S. Climate Change Science Program, 2008), based on a number of sources, report that "Extreme precipitation episodes (heavy downpours) have become more frequent and more intense in recent decades than at any other time in the historical record, and account for a larger percentage of total precipitation."

29. USGCRP (*Global Climate Change Impacts*, page 33), based on Kunkel et al., "Observed changes," pages 42-46. See also A. Dai, K. E. Trenberth, and T. Qian, "A global data set of Palmer Drought Severity Index for 1870-2002: Relationship with soil moisture and effects of surface warming" (*Journal of Hydrometeorology* 5(6):1117-1130, 2004).

30. USGCRP (*Global Climate Change Impacts*, page 82); see also CCSP SAP 4.3; J. S. Littell, D. McKenzie, D. L. Peterson, and A. L. Westerling, "Climate and wildfire area burned in western U. S. ecoprovinces, 1916-2003" (*Ecological Applications* 19[4]:1003-1021, 2009).

31. Probabilities are not attached to particular socioeconomic and emissions scenarios, but there is high confidence that realized future concentrations/radiative forcing resulting from anthropogenic activities will very likely fall within the range of estimates reflected in recent work establishing the new Representative Concentration Pathways. Note that this range does not include additional forcing possible from feedbacks.

32. Climate (or Earth system) models simulate the temporal evolution of the atmosphere, ocean, land surface, and other aspects of the climate system under certain assumptions and boundary conditions (such as, for example, different scenarios of future GHG emissions). These models are based on the fundamental laws of physics and chemistry, are calibrated and tested using observations of current and past climate change, and reflect current scientific understanding of relevant climate processes. However, certain features of the Earth system, such as clouds and the global carbon cycle, are either incompletely understood or cannot be fully resolved by current models, and so their effects must be approximated. As a result, for a given emissions scenario different models will predict somewhat different magnitude and details of future climate change.

33. Vulnerability is defined here as the degree to which a system is susceptible to, or unable to cope with, adverse effects of climate change. Vulnerability is a function of the character, magnitude, and rate of climate variation to which a system is exposed, its sensitivity, and its adaptive capacity.

34. IPCC (*Climate Change 2007 WG1*, Ch. 10). Note that IPCC chose not to assign a likelihood to this range because much of the variation comes from different assumptions about how the world will respond to climate change.

35. See, e.g., L. Tomassini, R. Knutti, G. Plattner, D. van Vuuren, T. Stocker, R. Howarth, and M. Borsuk, "Uncertainty and risk in climate projections for the 21st century: Comparing mitigation to non-intervention scenarios" (*Climatic Change* 103[3-4]:399-422, 2010, DOI 10.1007/s10584-009-9763-3). Also note that the scenarios on which these projections of future climate change are based do not actually reflect how policy interventions might influence future GHG emissions—they are rather illustrations of how climate change might evolve in the absence of global actions to reduce emissions. However, to the extent that the scenarios depicted in Figure 2.3 mirror emissions reductions that might be achieved through policy actions, they can be thought of as rough proxies for the temperature and

other climate changes expected under increasingly aggressive GHG emission reductions. As discussed below and in further detail in two of the ACC panel reports (NRC, *Advancing the Science* and NRC, *Limiting the Magnitude of Climate Change* [Washington, D.C.: National Academies Press, 2010]), recent scenario development efforts do include consideration of the socioeconomic, technological, and policy aspects of alternative GHG trajectories. These efforts have also focused on developing GHG trajectories in a more integrated and iterative manner with climate model projections and assessments of current and future climate impacts.(e.g., CCSP, *Scenarios of Greenhouse Gas Emissions and Atmospheric Concentrations*, Sub-report 2.1A of Synthesis and Assessment Product 2.1 by the U.S. Climate Change Science Program and the Subcommittee on Global Change Research, eds. L. Clarke, J. Edmonds, H. Jacoby, H. Pitcher, J. Reilly, and R. Richels [Washington, D.C.: Department of Energy, Office of Biological and Environmental Research, 2007]; L. Clarke, J. Edmonds, V. Krey, R. Richels, S. Rose, and M. Tavoni, "International climate policy architectures: Overview of the EMF 22 International Scenarios" [*Energy Economics* 31(Supplement 2):S64-S81, 2009], R. H. Moss, J. A. Edmonds, K. A. Hibbard, M. R. Manning, S. K. Rose, D. P. Van Vuuren, T. R. Carter, S. Emori, M. Kainuma, T. Kram, G. A. Meehl, J. F. B. Mitchell, N. Nakicenovic, K. Riahi, S. J. Smith, R. J. Stouffer, A. M. Thomson, J. P. Weyant, and T. J. Wilbanks, "The next generation of scenarios for climate change research and assessment" [*Nature* 463(7282):747-756, 2010]).

36. NRC, *Climate Stabilization Targets: Emissions, Concentrations, and Impacts over Decades to Millennia* (Washington, D.C.: National Academies Press, 2010).
37. Ibid.
38. IPCC (*Climate Change 2007 WG1*, Summary for Policymakers):"It is very likely [greater than 90% likelihood] that hot extremes, heat waves, and heavy precipitation events will continue to become more frequent."
39. R. J. Nicholls and A. Cazenave, "Sea-level rise and its impact on coastal zones" (*Science* 328[5985]:1517-1520, 2010, doi:10.1126/science.1185782); NRC, *Advancing the Science*, pages 191-193).
40. USGCRP (*Global Climate Change Impacts*, pages 62-63 and references therein).
41. S. Solomon, G.-K. Plattner, R. Knutti, and P. Friedlingstein, "Irreversible climate change due to carbon dioxide emissions" (*Proceedings of the National Academy of Sciences* 106:1707-1709, 2009).
42. IPCC (*Climate Change 2007 WG2*, Summary for Policymakers) and also S. C. Doney, V. J. Fabry, R. A. Feely, and J. A. Kleypas, "Ocean acidification: The other CO_2 problem" (*Annual Review of Marine Science* 1:169-192, 2009); L. Cao, and K. Caldeira, "Atmospheric CO_2 stabilization and ocean acidification" (*Geophysical Research Letters* 35:L19609, 2008); J. Silverman, B. Lazar, L. Cao, K. Caldeira, and J. Erez, "Coral reefs may start dissolving when atmospheric CO_2 doubles" (*Geophysical Research Letters* 36:L05606, 2009, doi:10.1029/2008GL036282).
43. USGCRP (*Global Climate Change Impacts*, pages 84-85)
44. Ocean acidification also poses significant risks to a wide range of other marine organisms (NRC, *Ocean Acidification*).
45. USGCRP (*Global Climate Change Impacts*, pages 47 and 83 and references therein).
46. Ibid., pages 71-78.
47. Ibid., (page 83), see also A. L. Westerling and B. P. Bryant, "Climate change and wildfire in California" (*Climatic Change* 87[1]:1-19, 2008, doi:10.1007/s10584-007-9363-z); NRC, *Stabilization Targets*.
48. NRC, *Ecological Impacts of Climate Change* (Washington, D.C.: National Academies Press, 2008); IPCC, *Climate Change 2007 WG2*, Ch. 4.
49. USGCRP (*Global Climate Change Impacts*, pages 89-98).
50. Center for Integrative Environmental Research, *The US Economic Impacts of Climate Change and the Costs of Inaction: A Review and Assessment by the Center for Integrative Environmental Research at the University of Maryland* (College Park, MD: University of Maryland, 2007, available at: *http:// www.cier.umd.edu/climateadaptation/*, accessed March 1, 2011); Congressional Budget Office (CBO), *Potential Impacts of Climate Change in the United States* (Washington, D.C.: Congressional Budget Office, 2009, available at *http://www.cbo.gov/ftpdocs/101xx/doc10107/05-04-ClimateChange_forWeb.pdf*, accessed March 1, 2011).

CHAPTER 3

1. For additional discussion, see Solomon et al., "Irreversible climate change" and NRC, *Stabilization Targets*.

2. It is estimated that 80-90 percent of the heating associated with GHG emissions over the past 50 years has gone into raising the temperature of the world's oceans (K. E. Trenberth and J. T. Fasullo, "Tracking Earth's energy" *Science* 328[5976]:316-317.2010). Also, it is estimated that even if atmospheric GHG concentrations could be immediately stabilized, an additional 0.1°C of warming per decade would be experienced over the next several decades (IPCC, *Climate Change 2007 WG1*).

3. See Solomon et al., "Irreversible climate change"; NRC, *Stabilization Targets*; NRC, *Ocean Acidification*; M. Meinshausen, N. Meinshausen, W. Hare, S. C. B. Raper, K. Frieler, R. Knutti, D. J. Frame, and M. R. Allen, "Greenhouse-gas emission targets for limiting global warming to 2 degrees Celsius" (*Nature* 458[7242]:1158-U1196, 2009, doi: 10.1038/nature08017).

4. See NRC, *Limiting the Magnitude*, p.114.

5. For critical analyses related to allocating emissions reduction responsibility, including views on how to implement the UNFCCC principle of "common but differentiated responsibilities," see S. Caney, "Cosmopolitan justice, responsibility, and global climate change" (*Leiden Journal of International Law* 18[4]:747-775, 2005) and E. A. Page, "Distributing the burdens of climate change." (*Environmental Politics* 17[4]:556-575, 2008). For a recent proposal for allocating responsibilities among individuals, see S. Chakravarty, A. Chikkatur, H. de Coninck, S. Pacala, R. Socolow, and M. Tavoni, "Sharing global CO_2 emission reductions among one billion high emitters" (*Proceedings of the National Academy of Sciences* 106[29]:11884-11888, 2009).

6. See NRC, *Limiting the Magnitude* for detailed discussion about the magnitude of these sorts of co-benefits.

7. Discussed further in NRC, *Advancing the Science*, Chapter 17, and NRC, *Limiting the Magnitude*, Chapter 2).

8. See NRC, *America's Climate Choices: Adapting to the Impacts of Climate Change*. Washington, D.C.: National Academies Press, 2010; and IPCC, *Climate Change 2007 WG2* for additional details.

9. See NRC, *Advancing the Science*, Chapter 17.

10. For instance, G. F. Nemet, T. Holloway, and P. Meier, "Implications of incorporating air-quality co-benefits into climate change policy making" (*Environmental Research Letters* 5, 2010) surveyed 37 studies of the economic benefit of air pollutant reductions that accompanied climate change mitigation efforts (given in dollars per ton of CO_2 avoided, in 2008 dollars). For developed countries, benefits ranged from $2-128 / tCO_2$. For developing counties, the range was $27-196 / tCO_2$. Developing countries generally have much higher levels of air pollution, and thus the incremental benefits of pollution mitigation are much greater.

11. T. Searchinger, R. Heimlich, R. A. Houghton, F. Dong, A. Elobeid, J. Fabiosa, S. Tokgoz, D. Hayes, and T. H. Yu, "Use of U.S. croplands for biofuels increases greenhouse gases through emissions from land-use change" (*Science* 319[5867]:1238-1240, 2008); D. A. Landis, M. M. Gardiner, W. van der Werf, and S. M. Swinton, "Increasing corn for biofuel production reduces biocontrol services in agricultural landscapes" (*Proceedings of the National Academy of Sciences* 105[51]:20552-20557, 2008).

12. This topic is addressed at length in NRC, *Informing Effective Decisions* and in NRC, *Facilitating Climate Change Responses: A Report of Two Workshops on Knowledge from the Social and Behavioral Sciences* (Washington, D.C.: National Academies Press, 2010), Chapter 1.

13. C. Keller, M. Siegrist, and H. Gutscher, "The role of the affect and availability heuristic in risk communication" (*Risk Analysis* 26:631-639, 2006); R. Hertwig, G. Barron, E. U. Weber, and I. Erev, "Decisions from experience and the effect of rare events" (*Psychological Science* 15:534-539 2004); and E. U. Weber and P. C. Stern, "Public understanding of climate change in the United States" (*American Psychologist*, 2011, in press).

14. A. Bostrom, M. G. Morgan, B. Fischhoff, and D. Read, "What do people know about global climate change? 1. Mental models" (*Risk Analysis* 14:959-970, 1994); T. W. Reynolds, A. Bostrom, D. Read, and M. G. Morgan, "Now what do people know about global climate change? Survey studies of educated laypeople" (*Risk Analysis* 30(10):1520-1538, 2010).

15. For instance, Sterman and Booth-Sweeney ("Understanding public complacency about climate change: Adults' mental models of climate change violate conservation of matter" [*Climatic Change* 80:213-238, 2007]) found that

63% of a sample of MIT graduate students believed that atmospheric CO_2 concentrations can be stabilized under a scenario where the amount of CO_2 emitted to the atmosphere exceeded the amount being removed from the atmosphere; A. Leiserowitz, N. Smith, and J. R Marlon, *Americans' Knowledge of Climate Change* (Yale Project on Climate Change Communication. New Haven, Connecticut: Yale University, 2010).

16. NRC, *Facilitating Climate Change Responses;* Weber and Stern, "Public understanding."
17. See, e.g., R. E. Dunlap and A. M. McCright, 2008. "Widening gap: Republican and Democratic views on climate change" (*Environment* 50:26-35, 2008); R. E. Dunlap and A. M. McCright, "Climate change denial: Sources, actors, and strategies." in *Routledge Handbook of Climate Change and Society*, ed. C. Lever-Tracy (New York: Routledge, 2010); M. Hulme, *Why We Disagree About Climate Change: Understanding Controversy, Inaction and Opportunity* (Cambridge, UK: Cambridge University Press, 2009); N. Oreskes and E. M. Conway, *Merchants of Doubt: How a Handful of Scientists Obscured the Truth on Issues from Tobacco Smoke to Global Warming.* (New York: Bloomsbury Press, 2010).
18. See, e.g., National Intelligence Council, *Global Trends 2025: A Transformed World* (Washington, D.C.: US Government Printing Office, 2008, available at: *http://www.dni.gov/nic/PDF_2025/2025_Global_Trends_Final_Report.pdf,* accessed March 3, 2011, 2008); CNA Corporation, *National Security and the Threat of Climate Change* (Alexandria, VA: The CNA Corporation, 2007, available at: *http://securityandclimate.cna.org,* accessed March 1, 2011).
19. J. Hansen, M. Sato, R. Ruedy, P. Kharecha, A. Lacis1, R. Miller, L. Nazarenko, K. Lo, G. A. Schmidt, G. Russell, I. Aleinov, S. Bauer, E. Baum, B. Cairns, V. Canuto, M. Chandler, Y. Cheng, A. Cohen, A. Del Genio, G. Faluvegi, E. Fleming, A. Friend, T. Hall, C. Jackman, M. Kelley, N. Y. Kiang, D. Koch, G. Labow, J. Lerner, S. Menon, T. Novakov, V. Oinas, Ja. Perlwitz, Ju. Perlwitz, D. Rind, A. Romanou, R. Schmunk, D. Shindell, P. Stone, S. Sun, D. Streets, N. Tausnev, D. Thresher, N. Unger, M. Yao, and S. Zhang, Dangerous human-made interference with climate: A GISS model study" (*Atmospheric Chemistry and Physics* 7:2287-2312, 2007); *Proceedings of the National Academy of Sciences*, Tipping Elements in Earth Systems Special Feature (*PNAS* 106[49]:20561, 2009).
20. M. G. Morgan, M. Kandlikar, J. Risbey, and H. Dowlatabadi, "Why conventional tools for policy analysis are often inadequate for problems of global change" (*Climatic Change* 41:271-281, 1999).

CHAPTER 4

1. See also NRC, *Informing Decisions in a Changing Climate* (Washington, D.C.: National Academies Press, 2009) and NRC, *Informing Effective Decisions*.
2. C. E. Lindblom, "The science of 'muddling through'" (*Public Administration Review* 19[2], 1959, available at *http://www.emerginghealthleaders.ca/resources/Lindblom-Muddling.pdf*, accessed March 3, 2011).
3. The United Nations Educational, Scientific and Cultural Organization (UNESCO) *The Precautionary Principle. World Commission on the Ethics of Scientific Knowledge and Technology* (Paris: UNESCO, 2005, available at *http://unesdoc.unesco.org/images/0013/001395/139578e.pdf*, accessed March 4, 2011) defines the precautionary principle as follows: Where there are threats of serious or irreversible damage, lack of full scientific certainty shall not be used as a reason for postponing cost effective measures to prevent environmental degradation.
4. C. R. Sunstein, *Laws of Fear: Beyond the Precautionary Principle* (New York: Cambridge University Press, 2005); R. J. Lempert and M. T. Collins, "Managing the risk of uncertain threshold responses: Comparison of robust, optimum, and precautionary approaches" (*Risk Analysis* 27[4]:1009-1026, 2007).
5. See, e.g., W. D. Nordhaus, *A Question of Balance. Weighing the Options on Global Warming Policies* (New Haven. CT: Yale University Press, 2008); R. S. J. Tol, "Equitable cost-benefit analysis of climate change" (*Ecological Economics* 36(1):71-85, 2001); N. Stern, *Stern Review on the Economics of Climate Change* (London, U.K.: H.M. Treasury, 2007).
6. See also NRC, *Limiting the Magnitude;* P. Watkiss, and T. Downing, "The social cost of carbon: Valuation estimates and their use in UK policy" (*Integrated Assessment Journal* 8[1]:85, 2008).
7. NRC (*Hidden Costs of Energy: Unpriced Consequences of Energy Production and Use* [Washington, D.C.: National Academies Press, 2009]) found that "depending on the [assumed] extent of future damages and the discount rate used for weighting future damages, the range of estimates of marginal global damages can vary by two orders of magnitude."

8. See NRC, *Limiting the Magnitude* and NRC, *America's Energy Future*.

9. Iterative risk management is sometimes used interchangeably with the term adaptive risk management. We chose to use "iterative," because in this report, "adaptive" is used in other contexts (e.g., in the context of adaptation to climate change impacts). Also, for the ecosystem management community, adaptive risk management is a term of art with a specific meaning that does not fully encompass the concepts being discussed here. See NRC, *Informing Effective Decisions* for further discussion and references on this topic.

10. IPCC, *Climate Change 2007 WG2*; World Bank, *Managing Climate Risk: Integrating Adaptation into World Bank Group Operations* (Washington, D.C.: World Bank Group, 2006, available at *http://siteresources.worldbank.org/GLOBAL ENVIRONMENTFACILITYGEFOPERATIONS/Resources/Publications-Presentations/GEFAdaptationAug06.pdf*, accessed March 17, 2011); United Nations Development Programme (UNDP), *A Climate Risk Management Approach to Disaster Reduction and Adaption to Climate Change* (Havana: UNDP and Harvard Medical School Center for Human Health and the Global Environment, 2002); Australian Greenhouse Office, *Climate Change Impacts and Risk Management: A Guide for Business and Government* (Canberra: Australian Greenhouse Office, 2006); UK Climate Change Risk Assessment (*http://archive.defra.gov.uk/environment/climate/adaptation/ccra/*).

11. NRC, *Informing Decisions in a Changing Climate, Adapting to the Impacts*, and *Informing Effective Decisions*; G. Yohe and R. Leichenko, "Adopting a risk-based approach," pp. 29-40 in *Climate Change Adaptation in New York City: Building a Risk Management Response*, New York City Panel on Climate Change 2010 Report (*Annals of the New York Academy of Sciences* 1196, 2010).

12. See NRC, *Advancing the Science* and *Limiting the Magnitude*.

13. e.g., H. Raiffa, *Decision Analysis: Introductory Lectures on Choice Under Uncertainty* (Reading, MA: Addison-Wesley, 1968); E. Crouch and R. Wilson, *Risk/Benefit Analysis* (Cambridge, MA: Ballinger, 1982); G. Suter, *Ecological Risk Analysis* (Boca Raton, FL: Lewis, 1993).

14. e.g., IPCC, *Climate Change 2007 WG2*, Chapter 19; J. B. Smith, S. H. Schneider, M. Oppenheimer, G .W. Yohe, W. Hare, M. D. Mastrandrea, A. Patwardhan, I. Burton, J. Corfee-Morloti, C.H.D. Magadza, H-M. Füssel, A. B. Pittock, A. Rahman, A. Suarez, and J-P van Ypersele. "Assessing dangerous climate change through an update of the Intergovernmental Panel on Climate Change (IPCC) 'reasons for concern.'" (*Proceedings of the National Academy of Sciences* 106[11]:4133-4137, 2009).

15. NRC, *America's Energy Future*; NRC, *Electricity from Renewable Resources: Status, Prospects, and Impediments* (Washington D.C.: National Academies Press, 2010).

16. Reviews of several experiences with emissions trading can be found in T. Tietenberg, "The tradable permits approach to protecting the commons: What have we learned?," pp. 197-232 in T*he Drama of the Commons*, ed. E. Ostrom, T. Dietz, N. Dolsak, P. Stern, S. Stonich, and E. Weber (Washington D.C.: National Academy Press, 2002); and T. Tietenberg, "The evolution of emissions trading," pp 42-58 in *Better Living Through Economics*, ed. J. J. Siegfried (Cambridge, MA.: Harvard University Press, 2010).

17. NRC, *America's Energy Future*.

18. NRC, *Hidden Costs of Energy*.

19. P. N. Leiby, *Estimating the Energy Security Benefits of Reduced U.S. Oil Imports* (Oak Ridge, TN: Oak Ridge National Laboratory, 2007).

20. L. Elbakidze and B. A. McCarl, Sequestration offsets versus direct emission reductions: Consideration of environmental co-effects (*Ecological Economics* 60[3]:564-571, 2007).

21. M. R Shammin and C. W. Bullard, "Impact of cap-and-trade policies for reducing greenhouse gas emissions on U.S. households" (*Ecological Economics* 68[8-9]:2432-2438, 2009).

22. See NRC, *Advancing the Science* and *Adapting to the Impacts*.

23. This difference is due primarily to the fact that adjusting standards generally requires a lengthy notice-and-comment administrative process and often entails litigation, whereas cap-and-trade systems can be designed to automatically adjust over time and keep costs within reasonable bounds.

24. For example multi-attribute utility analysis methods (R. L. Keeney and H. Raiffa, *Decisions with Multiple Objectives*. Second Edition (Cambridge, UK: Cambridge University Press, 1993).

CHAPTER 5

1. Copenhagen Accord (*http://unfccc.int/resource/docs/2009/cop15/eng/l07.pdf*); G-8 declaration (*http://www.g8ita-lia2009.it/static/G8_Allegato/MEF_Declarationl.pdf*).

2. e.g., see J. Hansen, M. Sato, P. Kharecha, D. Beerling, R. Berner, V. Masson-Delmotte, M. Pagani, M. Raymo, D. Royer, and J. Zachos, "Target atmospheric CO_2: Where Should humanity aim?" (*The Open Atmospheric Science Journal*, 2008[2]:217-223, 2008).

3. e.g., NRC, *Limiting the Magnitude* and *Advancing the Science*.

4. Meinshausen et al., "Greenhouse-gas emission targets"; M. R. Allen, D. J. Frame, C. Huntingford, C. D. Jones, J. A. Lowe, M. Meinshausen, and N. Meinshausen, "Warming caused by cumulative carbon emissions towards the trillionth tonne" (*Nature* 458:1163-1166, 2009); Wissenschaftlicher Beirat der Bundesregierung Globale Umweltveränderungen (WBGU, German Advisory Council on Global Change), *Solving the Climate Dilemma: The Budget Approach* (Berlin: WBGU, 2009).

5. e.g., see J. E. Aldy and R. N. Stavins, *Post-Kyoto International Climate Policy: Implementing Architectures for Agreement.* (New York: Cambridge University Press, 2010); J. B. Wiener and R. B. Stewart, *Reconstructing Climate Policy: Beyond Kyoto* (Washington, D. C.: American Enterprise Institute, 2003).

6. Clarke et al., "International climate policy architectures."

7. B. C. Murray, A. J. Sommer, B. Depro, B. L. Sohngen, B. A. McCarl, D. Gillig, B. De Angelo, and K. Andrasko. 2005. *Greenhouse Gas Mitigation Potential in US Forestry and Agriculture* (Washington, D.C.: Environmental Protection Agency, 2005); IPCC, *Climate Change 2007 WG1*; NRC, *Limiting the Magnitude*; NRC, *Advancing the Science*.

8. A Renewable Portfolio Standard requires electric utilities and other retail electric providers to supply a specified minimum amount of customer load with electricity from eligible renewable energy sources, with the goal of stimulating market and technology development and making renewable energy economically competitive with conventional forms of electric power. Such standards are in place in 29 states and the District of Columbia. Some have proposed "no-carbon" standards, which would include nuclear power as well as renewables.

9. NRC, *New Tools for Environmental Protection: Education, Information and Voluntary Measures* (Washington, D.C.: National Academy Press, 2002).

10. See, e.g., C. Fischer and R. G. Newell, "Environmental and technology policies for climate mitigation" (*Journal of Environmental Economics and Management* 55(2):142-162, 2008); T. H. Tietenberg, *Emissions Trading: Principles and Practice.* (Washington, D.C.: Resources for the Future, 2006); NRC, *Limiting the Magnitude*.

11. Trades of this kind regularly occur within the European Union's Emissions Trading System. These trades require only the measurement of actual emissions, not the estimation of what emissions would have been absent the trade. On financial flows within the EU ETS, see Aldy and Stavins, *Post-Kyoto*.

12. EMF22: Clarke et al., "International climate policy architectures;" A. A. Fawcett, K. V. Calvin, F. C. De La Chesnaye, J. M. Reilly, and J. P. Weyant, "Overview of EMF 22 U.S. transition scenarios" (*Energy Economics* 31[Supplement 2]:S198-S211, 2009).

13. A concept within economic theory wherein the allocation of goods and services by a free market is not efficient.

14. See NRC, *Limiting the Magnitude* for detailed discussion.

15. Some examples discussed in M. A. Brown and S. Chandler, "Governing confusion: How statutes, fiscal policy, and regulations impede clean energy technologies" (*Stanford Law and Policy Review* 19[3]:472-509, 2008, available at *http://slpr.stanford.edu/previous/Volume19.html#Issue3*, accessed March 1, 2011): Ten states have no statewide energy codes for residential construction or have codes that predate 1998; seven states do not have net metering for distributed power generation; 41 states have not decoupled electric utility profits from electricity sales; and all states ban private electric wires crossing public streets, which forces would-be power entrepreneurs to use their competitors' wires.

16. See M. P. Vandenbergh, P. C. Stern, G. T. Gardner, T. Dietz, and J. M. Gilligan, "Implementing the behavioral wedge: Designing and adopting effective carbon emissions reduction programs" (*Environmental Law Review* 40:10547-10554, 2010); P. C. Stern, G. T Gardner, M. P Vandenbergh, T. Dietz, and J. M Gilligan, "Design principles for carbon emissions reduction programs" (*Environmental Science and Technology* 44[13]:4847-4848, 2010).

17. As compared to a comprehensive carbon price, a renewable portfolio standard does not provide incentives for efficiency in energy use, and its support of only selected technologies is unlikely to produce least-cost outcomes. Similarly, a cap-and-trade system covering only some sectors would minimize costs within but not across sectors. A recent study of twenty CO_2 reduction policies (Resources for the Future and National Energy Policy Institute (RFF/NEPI), *Toward a New National Energy Policy: Assessing the Options* (Washington, D.C.: Resources for the Future, 2010, available at *http://www.rff.org/toward-a-new-energy-policy*, accessed March 4, 2011) suggests that through 2030, some alternatives to a comprehensive pricing systems (such as a cap-and-trade policy excluding transportation) are reasonably cost-effective.

18. As of February 2011, several bills have been introduced in Congress to delay or block the EPA from moving ahead with implementation of any new rules for regulating CO_2 emissions (e.g., see: T. Tracy, "Greenhouse-gas rules targeted by lawmakers" [*Wall Street Journal*, January 6, 2011]).

19. RFF/NEPI, *Toward a New National Energy Policy*.

20. See NRC, *Advancing the Science*, Chapters 4 and 6.

21. Tribal communities are listed as a distinct category because Native American communities that live on established reservations have unique vulnerabilities, given their limited relocation options.

22. See NRC, *Adapting to the Impacts* for additional details and discussion.

23. See NRC, *Adapting to the Impacts* for numerous examples of such policies.

24. See NRC, *Adapting to the Impacts* for additional examples of currently existing maladaptive policies and practices.

25. See NRC, *Adapting to the Impacts* and references therein.

26. See NRC, *Understanding Risk, Public Participation, and Informing Decisions*.

27. NRC, *Verifying Greenhouse Gas Emissions*.

28. See NRC, *Public Participation, Informing Decisions, and Informing Effective Decisions*.

America's Climate Choices: Membership Lists

COMMITTEE ON AMERICA'S CLIMATE CHOICES

ALBERT CARNESALE (Chair), University of California, Los Angeles
WILLIAM CHAMEIDES (Vice Chair), Duke University, Durham, North Carolina
DONALD F. BOESCH, University of Maryland Center for Environmental Science, Cambridge
MARILYN A. BROWN, Georgia Institute of Technology, Atlanta
JONATHAN CANNON, University of Virginia, Charlottesville
THOMAS DIETZ, Michigan State University, East Lansing
GEORGE C. EADS, Charles River Associates, Washington, D.C.
ROBERT W. FRI, Resources for the Future, Washington, D.C.
JAMES E. GERINGER, Environmental Systems Research Institute, Cheyenne, Wyoming
DENNIS L. HARTMANN, University of Washington, Seattle
CHARLES O. HOLLIDAY, JR., DuPont, Wilmington, Delaware
KATHARINE L. JACOBS,* Arizona Water Institute, Tucson
THOMAS KARL,* National Oceanic and Atmospheric Administration, Asheville, North Carolina
DIANA M. LIVERMAN, University of Arizona, Tuscon, and University of Oxford, United Kingdom
PAMELA A. MATSON, Stanford University, California
PETER H. RAVEN, Missouri Botanical Garden, St. Louis
RICHARD SCHMALENSEE, Massachusetts Institute of Technology, Cambridge
PHILIP R. SHARP, Resources for the Future, Washington, D.C.
PEGGY M. SHEPARD, WE ACT for Environmental Justice, New York, New York
ROBERT H. SOCOLOW, Princeton University, New Jersey
SUSAN SOLOMON, National Oceanic and Atmospheric Administration, Boulder, Colorado
BJORN STIGSON, World Business Council for Sustainable Development, Geneva, Switzerland

Asterisks (*) denote members who resigned during the study process

THOMAS J. WILBANKS, Oak Ridge National Laboratory, Tennessee
PETER ZANDAN, Public Strategies, Inc., Austin, Texas

PANEL ON LIMITING THE MAGNITUDE OF FUTURE CLIMATE CHANGE

ROBERT W. FRI (Chair), Resources for the Future, Washington, D.C.
MARILYN A. BROWN (Vice Chair), Georgia Institute of Technology, Atlanta
DOUG ARENT, National Renewable Energy Laboratory, Golden, Colorado
ANN CARLSON, University of California, Los Angeles
MAJORA CARTER, Majora Carter Group, LLC, Bronx, New York
LEON CLARKE, Joint Global Change Research Institute (Pacific Northwest National
    Laboratory/University of Maryland), College Park, Maryland
FRANCISCO DE LA CHESNAYE, Electric Power Research Institute, Washington, D.C.
GEORGE C. EADS, Charles River Associates, Washington, D.C.
GENEVIEVE GIULIANO, University of Southern California, Los Angeles
ANDREW J. HOFFMAN, University of Michigan, Ann Arbor
ROBERT O. KEOHANE, Princeton University, New Jersey
LOREN LUTZENHISER, Portland State University, Oregon
BRUCE MCCARL, Texas A&M University, College Station
MACK MCFARLAND, DuPont, Wilmington, Delaware
MARY D. NICHOLS, California Air Resources Board, Sacramento
EDWARD S. RUBIN, Carnegie Mellon University, Pittsburgh, Pennsylvania
THOMAS H. TIETENBERG, Colby College (retired), Waterville, Maine
JAMES A. TRAINHAM, RTI International, Research Triangle Park, North Carolina

PANEL ON ADAPTING TO THE IMPACTS OF CLIMATE CHANGE

KATHARINE L. JACOBS* (Chair, through January 3, 2010), University of Arizona, Tucson
THOMAS J. WILBANKS (Chair), Oak Ridge National Laboratory, Tennessee
BRUCE P. BAUGHMAN, IEM, Inc., Alabaster, Alabama
ROBERT BEACHY,* Donald Danforth Plant Sciences Center, Saint Louis, Missouri
GEORGES C. BENJAMIN, American Public Health Association, Washington, D.C.
JAMES L. BUIZER, Arizona State University, Tempe
F. STUART CHAPIN III, University of Alaska, Fairbanks
W. PETER CHERRY, Science Applications International Corporation, Ann Arbor,
    Michigan
BRAXTON DAVIS, South Carolina Department of Health and Environmental Control,
    Charleston
KRISTIE L. EBI, IPCC Technical Support Unit WGII, Stanford, California

JEREMY HARRIS, Sustainable Cities Institute, Honolulu, Hawaii
ROBERT W. KATES, Independent Scholar, Bangor, Maine
HOWARD C. KUNREUTHER, University of Pennsylvania Wharton School of Business, Philadelphia
LINDA O. MEARNS, National Center for Atmospheric Research, Boulder
PHILIP MOTE, Oregon State University, Corvallis
ANDREW A. ROSENBERG, Conservation International, Arlington, Virginia
HENRY G. SCHWARTZ, JR., Jacobs Civil (retired), Saint Louis, Missouri
JOEL B. SMITH, Stratus Consulting, Inc., Boulder, Colorado
GARY W. YOHE, Wesleyan University, Middletown, Connecticut

PANEL ON ADVANCING THE SCIENCE OF CLIMATE CHANGE

PAMELA A. MATSON (Chair), Stanford University, California
THOMAS DIETZ (Vice Chair), Michigan State University, East Lansing
WALEED ABDALATI, University of Colorado at Boulder
ANTONIO J. BUSALACCHI, JR., University of Maryland, College Park
KEN CALDEIRA, Carnegie Institution of Washington, Stanford, California
ROBERT W. CORELL, H. John Heinz III Center for Science, Economics and the Environment, Washington, D.C.
RUTH S. DEFRIES, Columbia University, New York, New York
INEZ Y. FUNG, University of California, Berkeley
STEVEN GAINES, University of California, Santa Barbara
GEORGE M. HORNBERGER, Vanderbilt University, Nashville, Tennessee
MARIA CARMEN LEMOS, University of Michigan, Ann Arbor
SUSANNE C. MOSER, Susanne Moser Research & Consulting, Santa Cruz, California
RICHARD H. MOSS, Joint Global Change Research Institute (Pacific Northwest National Laboratory/University of Maryland), College Park, Maryland
EDWARD A. PARSON, University of Michigan, Ann Arbor
A. R. RAVISHANKARA, National Oceanic and Atmospheric Administration, Boulder, Colorado
RAYMOND W. SCHMITT, Woods Hole Oceanographic Institution, Massachusetts
B. L. TURNER II, Arizona State University, Tempe
WARREN M. WASHINGTON, National Center for Atmospheric Research, Boulder, Colorado
JOHN P. WEYANT, Stanford University, California
DAVID A. WHELAN, The Boeing Company, Seal Beach, California

PANEL ON INFORMING EFFECTIVE DECISIONS AND ACTIONS RELATED TO CLIMATE CHANGE

DIANA LIVERMAN (Co-chair), University of Arizona, Tucson

PETER RAVEN (Co-chair), Missouri Botanical Garden, Saint Louis

DANIEL BARSTOW, Challenger Center for Space Science Education, Alexandria, Virginia

ROSINA M. BIERBAUM, University of Michigan, Ann Arbor

DANIEL W. BROMLEY, University of Wisconsin-Madison

ANTHONY LEISEROWITZ, Yale University

ROBERT J. LEMPERT, The RAND Corporation, Santa Monica, California

JIM LOPEZ,* King County, Washington

EDWARD L. MILES, University of Washington, Seattle

BERRIEN MOORE III, Climate Central, Princeton, New Jersey

MARK D. NEWTON, Dell, Inc., Round Rock, Texas

VENKATACHALAM RAMASWAMY, National Oceanic and Atmospheric Administration, Princeton, New Jersey

RICHARD RICHELS, Electric Power Research Institute, Inc., Washington, D.C.

DOUGLAS P. SCOTT, Illinois Environmental Protection Agency, Springfield

KATHLEEN J. TIERNEY, University of Colorado at Boulder

CHRIS WALKER, The Carbon Trust LLC, New York, New York

SHARI T. WILSON, Maryland Department of the Environment, Baltimore

Committee on America's Climate Choices Member Biographical Sketches

Dr. Albert Carnesale (NAE) (*Chair*) is Chancellor Emeritus and Professor at the University of California, Los Angeles (UCLA). He was Chancellor of the University from 1997 through 2006 and now serves as Professor of Public Policy and of Mechanical and Aerospace Engineering. His research and teaching focus on public policy issues having substantial scientific and technological dimensions, and he is the author or co-author of six books and more than 100 articles on a wide range of subjects, including national security strategy, arms control, nuclear proliferation, the effects of technological change on foreign and defense policy, domestic and international energy issues, and higher education. He is a member of the Secretary of Energy's Blue Ribbon Commission on America's Nuclear Future, the Mission Committees of the Lawrence Livermore National Laboratory and the Los Alamos National Laboratory, the World Economic Forum's Global Agenda Council on Weapons of Mass Destruction, the Board of Directors of Harvard University's Belfer Center for Science and International Affairs, and the Advisory Board of the RAND Corporation's Center for Global Risk and Security; and he chaired the National Academies Committees on Conventional Prompt Global Strike Capability and on Nuclear Forensics. Prior to joining UCLA, he was at Harvard for 23 years, serving as Lucius N. Littauer Professor of Public Policy and Administration, Dean of the John F. Kennedy School of Government, and Provost of the University. Before that, he served in government and in industry. Dr. Carnesale holds bachelor's and master's degrees in mechanical engineering and a Ph.D. in nuclear engineering. He is a fellow of the American Academy of Arts and Sciences and of the American Association for the Advancement of Science and is a member of the Council on Foreign Relations.

Dr. William L. Chameides (NAS) (*Vice Chair*) is the Dean of the Nicholas School of the Environment at Duke University, a position he has held since 2007. Prior to joining Duke he spent 3 years as the chief scientist of the Environmental Defense Fund, following more than 30 years in academia as a professor, researcher, teacher, and mentor. Chameides' research focuses on the atmospheric sciences, elucidating the causes of and remedies for global, regional, and urban environmental change and identifying pathways toward a more sustainable future. Specifically his research helped lay the groundwork for our understanding of the photochemistry of the lower atmosphere, elucidated the importance of nitrogen oxides emission controls in the mitigation of urban and regional photochemical smog, and the impact of regional air pollution on global food production. He has led two major, multi-institutional research projects: the Southern Oxidants Study, a research program focused on understanding the causes and remedies for air pollution in the southern United States; and CHINA-MAP, an international research program studying the effects of environmental change on agriculture in China. He is a member of the National Academy of Sciences, fellow of the American Geophysical Union and recipient of the American Geophysical Union's MacElwane Award. Chameides has served on numerous national and international committees and task forces and in recognition was named a National Associate of the National Academies for "extraordinary service."

Dr. Donald F. Boesch is a Professor of Marine Science and President of the University of Maryland Center for Environmental Science. He also serves as I Vice Chancellor for Environmental Sustainability of the University System of Maryland. Boesch is a biological oceanographer who has conducted research in coastal and continental shelf environments along the Atlantic Coast and in the Gulf of Mexico, eastern Australia and the East China Sea. He has published two books and more than 90 papers on estuarine and continental shelf ecosystems, oil pollution, nutrient over-enrichment, environmental assessment and monitoring, and science policy. Presently his research focuses on the use of science in ecosystem management and climate change adaptation. He was a contributing author to the U.S. Global Change Research Program report *Global Climate Change Impacts in the United States*. He was appointed by President Obama as a member of the National Commission on the BP Deepwater Horizon Oil Spill and Offshore Drilling. A native of New Orleans, Boesch received his B.S. from Tulane University and Ph.D. from the College of William & Mary. He was a Fulbright Postdoctoral Fellow at the University of Queensland and subsequently served on the faculty of the Virginia Institute of Marine Science. In 1980 he became the first Executive Director of the Louisiana Universities Marine Consortium, where he was also a Professor of Marine Science at Louisiana State University. He assumed his present position in Maryland in 1990.

Dr. Marilyn A. Brown is an endowed Professor of Energy Policy in the School of Public Policy at the Georgia Institute of Technology, which she joined in 2006 after a distinguished career at the U.S. Department of Energy's Oak Ridge National Laboratory. At ORNL, she held various leadership positions and co-led the report, Scenarios for a Clean Energy Future, which remains a cornerstone of engineering-economic analysis of low-carbon energy options for the United States. Her research interests encompass the design of energy and climate policies, issues surrounding the commercialization of new technologies, and methods for evaluating sustainable energy programs and policies. Dr. Brown has authored more than 250 publications including a recently published book on *Energy and American Society: Thirteen Myths* and a forthcoming book, *Climate Change and Energy Security*. Dr. Brown has been an expert witness in hearings before Committees of both the U.S. House of Representatives and the U.S. Senate, and she participates on several National Academies Boards and Committees. Dr. Brown has a Ph.D. in Geography from the Ohio State University, a master's degree in Resource Planning from the University of Massachusetts and is a Certified Energy Manager.

Mr. Jonathan Cannon is Professor of Law and Director of the University of Virginia Law School's Environmental and Land Use Law Program. Prior to joining the Law School faculty in 1998, he was at the Environmental Protection Agency (EPA), where he served as General Counsel from 1995 to 1998 and as Assistant Administrator for Administration and Resources Management from 1992 to 1995. He also held senior management positions at EPA from 1986 to 1990. Prior to his work with the EPA, Cannon was in the private practice of environmental law. He has written widely in environmental law and policy, with an emphasis on institutional design and adaptive management. He received his J.D. from University of Pennsylvania Law School and his B.A. from Williams College.

Dr. Thomas Dietz is Assistant Vice President for Environmental Research, Professor of Sociology and Environmental Science and Policy at Michigan State University. His current research examines the human driving forces of environmental change, environmental values and the interplay between science and democracy in environmental issues. Dietz is also an active participant in the Ecological and Cultural Change Studies Group at MSU. He is a Fellow of the American Association for the Advancement of Science and has been awarded the Sustainability Science Award of the Ecological Society of America, the Distinguished Contribution Award of the American Sociological Association Section on Environment, Technology and Society, and the Outstanding Publication Award, also from the American Sociological Association Section on Environment, Technology and Society. He has served on numerous National Academies' panels and committees and chaired the Committee on the Human Dimensions of Global Change and the Panel on Public Participation in Environmental Assessment and Decision Making.. He holds a Bachelor of General Studies degree from Kent State and a PhD in Ecology from the University of California at Davis.

Dr. George C. Eads is a Senior Consultant of Charles River Associates (CRA). Prior to joining CRA in 1995, he held several positions at General Motors (GM) Corporation, including Vice President and Chief Economist; Vice President, Worldwide Economic and Market Analysis Staff; and Vice President, Product Planning and Economics Staff. Before joining GM, Dr. Eads was Dean of the School of Public Affairs at the University of Maryland, College Park, where he also was a Professor. Before that, he served as a Member of President Carter's Council of Economic Advisors. He has been involved in numerous projects concerning transport and energy. In 1994 and 1995, he was a member of President Clinton's policy dialogue on reducing greenhouse gas emissions from personal motor vehicles. He co-authored the World Energy Council's 1998 Report, *Global Transport and Energy Development—The Scope for Change.* Over the past 4 years, Dr. Eads devoted most of his time to the World Business Council for Sustainable Development's Sustainable Mobility Project, a project funded and carried out by 12 leading international automotive and energy companies. Dr. Eads is a member of the Presidents' Circle at the National Academies. He is an at-large Director of the National Bureau of Economic Research. He received a Ph.D. degree in economics from Yale University. He is currently participating in the Transportation Research Board (TRB) study on "Potential Greenhouse Gas Reductions from Transportation" and recently completed service on the TRB study on "Climate Change and U.S. Transportation."

Mr. Robert W. Fri is a visiting scholar and senior fellow emeritus at Resources for the Future, a nonprofit organization that studies natural resource and environmental issues. He has served as director of the National Museum of Natural History, president of Resources for the Future, and deputy administrator of both the Environmental Protection Agency and the Energy Research and Development Administration. Fri has been director of American Electric Power Company; vice-chair and a director of the Electric Power Research Institute; a trustee and vice-chair of Science Service, Inc.; and a member of the National Petroleum Council. He is active with the National Academies, where he is a National Associate, vice-chair of the Board on Energy and Environmental Systems, and a member of the Advisory Board of the Marion E. Koshland Science Museum. He has chaired studies for the National Research Council on the health standards for the Yucca Mountain repository and on estimating the benefits of applied research programs at the Department of Energy. He currently chairs a study to evaluate the nuclear energy research program at DOE. Fri received his B.A. in physics from Rice University and his M.B.A. from Harvard University and is a member of Phi Beta Kappa and Sigma Xi.

The Honorable James E. Geringer received a B.S. in Mechanical Engineering from Kansas State University, then spent 10 years active and 12 years reserve service in the U.S. Air Force working on unmanned space programs for both the U.S. Air Force and NASA. Upon leaving active duty, he served as contract administrator for the construction of a 1,700 megawatt coal-fired electric power generation plant near Wheatland, Wyoming, then took up agricultural pursuits along with serving in the Wyoming Legislature from 1983 to 1994, including 6 years each in the House and the Senate. Geringer served two terms as Wyoming Governor. While in office, he chaired the

Western Governors' Association, the Education Commission of the States and served on a variety of national and regional education and technology initiatives. He served on the Mapping Sciences Committee under the National Research Council; Community Resilience Committee under Oak Ridge National Laboratories; Western Interstate Energy Board ; Vice-Chair of the Association of Governing Boards for Colleges and Universities; Operation Public Education; the Board of Governors of the Park City Center for Public Policy; Board member of NatureServe and, co-chair of the Policy Consensus Initiative. He is the current Chair of the Board of Trustees, Western Governors University. Jim joined Environmental Systems Research Institute (ESRI) in the summer of 2003 as Director of Policy and Public Sector Strategies to work with senior elected and corporate officials on how to use geospatial technology for place-based decisions in business and government.

Dr. Dennis L. Hartmann is currently Interim Dean of the College of the Environment, Professor in the Department of Atmospheric Sciences, Senior Fellow and Council Member of the Joint Institute for the Study of the Atmosphere and Ocean at the University of Washington. His research interests include dynamics of the atmosphere, atmosphere-ocean interaction, and climate change. His current research includes the study of climate feedback processes involving clouds and water vapor, which is approached using remote sensing data, in situ data and models, and attempts to take into account radiative, dynamical, and cloud-physical processes. Dr. Hartmann is a fellow of the American Meteorological Society, the American Geophysical Union, and the American Association for the Advancement of Science. He has served on numerous advisory, editorial and review boards for NSF, NASA, and NOAA and on multiple NRC committees, including the Committee on Climate Change Feedbacks (chair), Climate Research Committee, and Committee on Earth Sciences. He currently serves on the Board of Reviewing Editors for the magazine *Science* and is co-editor of the International Geophysics Series of Academic Press. Dr. Hartmann received his Ph.D. in Geophysical Fluid Dynamics from Princeton University.

Mr. Charles O. Holliday, Jr. (NAE) is chairman of the board of directors of Bank of America. He has served as a director since September 2009. He is the former chairman of the board of directors of E.I. du Pont de Nemours and Co., a position he had held for approximately 10 years. He served as chief executive officer of DuPont from 1998 until 2008. He joined DuPont in 1970 as an engineer and held various positions throughout his tenure. Since 2007, Holliday has served as a member of the board of directors of Deere & Co. and as a member of the board's audit and corporate governance committees. He is chairman emeritus of Catalyst, a leading nonprofit organization dedicated to expanding opportunities for women and business, and chairman emeritus of the board of the U.S. Council on Competitiveness, a nonpartisan, nongovernmental organization working to ensure U.S. prosperity. Holliday is a founding member of the International Business Council and a member of the National Academy of Engineering. He also previously served as chairman of the following organizations: the Business Roundtable's Task Force for Environment, Technology and Economy, the World Business Council for Sustainable Development, The Business Council, and the Society of Chemical Industry—American Section. He received a bachelor's degree in industrial engineering from the University of Tennessee and received honorary doctorates from Polytechnic University in Brooklyn, New York, and from Washington College in Chestertown, Maryland.

Dr. Diana M. Liverman holds joint appointments between Oxford University (as Senior Research Fellow in the Environmental Change Institute—ECI) and the University of Arizona (where she co-directs the Institute of the Environment). Her research has focused on the human dimensions of global environmental change, including climate impacts, governance, and policy; climate and development; and the political ecology of environment, land use, and development in Latin America. She has current projects on climate vulnerability and adaptation, climate impacts on food systems, and carbon offsets and has interest in connecting research to stakeholders and climate science to the arts and creative sector. She has led or coordinated major research programs for the Tyndall Center for Climate Change, the James Martin 21st Century School at Oxford, the Global Environmental Change and Food Systems project (GECAFS), the UK Climate Impacts Program, and the Climate Assessment for the Southwest (CLIMAS). Her advisory roles have included the NRC Committee on the Human Dimensions of Global Environmental Change (chair) and the scientific advisory committees for the InterAmerican Institute (IAI) for Global Change (co-chair). She has a B.A. in Geography from University College London, an M.A. from the University of Toronto, and a Ph.D. from UCLA.

Dr. Pamela A. Matson (NAS) is Chester Naramore Dean of the School of Earth Sciences at Stanford University. She is also the Richard and Rhoda Goldman Professor of Environmental Studies and senior fellow in the Woods Institute of Environment and Sustainability. Her research focuses on biogeochemical cycling and biosphere-atmosphere interactions in tropical forests and agricultural systems. Together with hydrologists, atmospheric scientists, economists, and agronomists, Matson analyzes the economic drivers and environmental consequences of land use and resource use decisions in developing world agricultural and natural ecosystems, with the objective of identifying practices that are economically and environmentally sustainable. With her students, she also evaluates the response of tropical forests to nitrogen deposition and climate changes. Matson joined the Stanford faculty in 1997, following positions as professor at UC Berkeley and research scientist at NASA. She is a past President of the Ecological Society of America, currently serves on the board of trustees of the World Wildlife Fund, and until recently was the chair of the National Academies' Roundtable on Science and Technology for Sustainability. She was elected to the American Academy of Arts and Sciences in 1992 and to the National Academy of Sciences in 1994. In 1995, Dr. Matson was selected as a MacArthur Fellow and in 1997 was elected a Fellow of the American Association for the Advancement of Science. In 2002 she was named the Burton and Deedee McMurtry University Fellow in Undergraduate Education at Stanford. She earned her B.S. at the University of Wisconsin—Eau Claire, M.S. at Indiana University, and Ph.D. at Oregon State University.

Dr. Peter H. Raven (NAS), President of the Missouri Botanical Garden, is one of the world's leading botanists and advocates of conservation and biodiversity. He received the National Medal of Science, the highest award for scientific accomplishment in the United States in December 2000. Raven has also received numerous other prizes and awards, including the Society for Conservation Biology Distinguished Service Award and the Peter H. Raven Award for Scientific Outreach, which was created to honor him. He also received the prestigious International Prize for Biology from the government of Japan; Environmental Prize of the Institute de la Vie; Volvo Environment Prize; the Tyler Prize for Environmental Achievement, the Sasakawa Environment Prize, and has held Guggenheim and John D. and Catherine T. MacArthur Foundation Fellowships. Described by *Time* magazine as a "Hero for the Planet," Raven champions research around the world to preserve endangered plants and is a leading advocate for conservation and a sustainable environment. For three decades Raven has headed the Missouri Botanical Garden, an institution he nurtured to a world-class center for botanical research, education, and horticulture display. He is also the Engleman Professor of Botany at Washington University in St. Louis, Chairman of the National Geographic Society's Committee for Research and Exploration, and previously served as President of the American Association for the Advancement of Science and as a member of the President's Committee of Advisors on Science and Technology. He served for 12 years as Home Secretary of the National Academy of Sciences, is a member of the academies of science in Argentina, China, India, Italy, Russia, and several other countries; belongs to the Pontifical Academy of Sciences, and was inducted into the American Academy of Achievement. He was first Chair of the U. S. Civilian Research and Development Foundation, a government-established organization that funds joint research with the independent countries of the former Soviet Union. Raven received his Ph.D. from the University of California, Los Angeles, in 1960 after completing his undergraduate work at the University of California, Berkeley. He has received honorary degrees from universities in this country and throughout the world.

Dr. Richard Schmalensee is the Howard W. Johnson Professor of Economics and Management at the Massachusetts Institute of Technology (MIT) and Director of the MIT Center for Energy and Environmental Policy Research. He served as the John C. Head III Dean of the MIT Sloan School of Management from 1998 through 2007. He was a Member of the President's Council of Economic Advisers from 1989 through 1991. Professor Schmalensee is the author or co-author of 11 books and more than 110 published articles, and he is co-editor of volumes 1 and 2 of the *Handbook of Industrial Organization*. His research has centered on industrial organization economics and its application to managerial and public policy issues, with particular emphasis on antitrust, regulatory, energy, and environmental policies. Professor Schmalensee is a Fellow of the Econometric Society and the American Academy of Arts and Sciences, and a Research Associate of the National Bureau of Economic Research. He has served as a member of the National Commission on Energy Policy and of the Executive Committee of the American Economic Association and as a director of the International Securities Exchange and other corporations. He is currently a director of the International Data Group and of Resources for the Future. He received his S.B. and Ph.D. in Economics at MIT.

Dr. Philip R. Sharp became President of Resources for the Future on September 1, 2005. His career in public service includes ten terms as a member of the U.S. House of Representatives from Indiana, beginning in 1975. He was a driving force behind the Energy Policy Act of 1992. He also helped to develop a critical part of the 1990 Clean Air Act Amendments, providing for a market-based emissions allowance trading system. After leaving Congress, he served on the faculty of the John F. Kennedy School of Government and the Institute of Politics at Harvard University from 1995 to 2005. Sharp was Congressional chair of the National Commission on Energy Policy (2004), the National Research Council's Committee on Effectiveness and Impact of Corporate Average Fuel Economy (CAFE) Standards (2001), and chair of the Secretary of Energy's Electric Systems Reliability Task Force (1998). Sharp is co-chair of the Energy Board of the Keystone Center and serves on the Board of Directors of the Duke Energy Corporation and the Energy Foundation. He is also a member of the Cummins Science and Technology Advisory Council and serves on the Advisory Board of the Institute of Nuclear Power Operations (INPO) and on the MIT Energy Initiative External Advisory Board. He served on the Board of Directors of the Cinergy Corporation from 1995–2006, on the Board of the Electric Power Research Institute from 2002–2006, and on the National Research Council's Board of Energy and Environmental Systems (BEES) from 2001–2007. In addition, he chaired advisory committees for the Massachusetts Institute of Technology studies on the future of nuclear power and the future of coal. Before accepting the RFF presidency, Sharp was senior policy advisor to the Washington law firm of Van Ness Feldman, and a senior advisor to the Cambridge economic analysis firm of Lexecon/FTI. Prior to his service in Congress, Sharp taught political science at Ball State University from 1969 to 1974. Sharp graduated cum laude from Georgetown University's School of Foreign Service in 1964 and received his Ph.D. in government from Georgetown in 1974.

Ms. Peggy M. Shepard is executive director and co-founder of WE ACT for Environmental Justice. Founded in 1988, WE ACT was New York's first environmental justice organization created to improve environmental health and quality of life in communities of color. She is the recipient of numerous awards for her leadership and advocacy, including the 10th Annual Heinz Award for the Environment and the 2008 Jane Jacobs Medal for Lifetime Achievement. She is a former Democratic District Leader, who represented West Harlem from 1985 to April 1993, and served as President of the National Women's Political Caucus-Manhattan from 1993–1997. From January 2001–2003, Ms Shepard served as the first female chair of the National Environmental Justice Advisory Council (NEJAC) to the U.S. Environmental Protection Agency, and is co-chair of the Northeast Environmental Justice Network. She is a former member of the National Advisory Environmental Health Sciences Council of the National Institutes of Health and a member of the Environmental Justice Advisory Committee to the NYS Department of Environmental Conservation. Ms. Shephard is a former journalist and was a reporter for *The Indianapolis News*, a copy editor for *The San Juan Star*, and a researcher for Time-Life Books. She has served as an editor at *Redbook*, *Essence*, and *Black Enterprise* magazines. Ms. Shepard began a career in government as a speechwriter for the New York State Division of Housing & Community Renewal and Director of Public Information for Rent Administration. She served as the Women's Outreach Coordinator for the New York City Comptroller's Office. Ms. Shepard is a board member of the national and NYS Leagues of Conservation Voters, Environmental Defense, NY Earth Day, Citizen Action of NY, the Children's Environmental Health Network, and Healthy Schools Network, Inc. She is an advisory board member of the Bellevue Occupational and Environmental Medicine Clinic, the Harlem Center for Health Promotion and Disease Prevention, and Mt. Sinai's Children's Environmental Health Center. She is a graduate of Howard University and Solebury and Newtown Friends Schools.

Dr. Robert H. Socolow is a Professor of Mechanical and Aerospace Engineering at Princeton University, where he teaches in both the School of Engineering and Applied Science, and the Woodrow Wilson School of Public and International Affairs and co-director of the University's Carbon Mitigation Initiative. He was the Director of the University's Center for Energy and Environmental Studies from 1979 to 1997. His current research focuses on the characteristics of a global energy system that would be responsive to global and local environmental and security constraints. His specific areas of interest include the capture of carbon dioxide from fossil fuels and its storage in geological formations, nuclear power, energy efficiency in buildings, and the accelerated deployment of advanced technologies in developing countries. He was editor of *Annual Review of Energy and the Environment* from 1992 to 2002. He is a National Associate of the U.S. National Academies and a Fellow of the American Physical Society and the American Association for the Advancement of Science. He was awarded the 2003 Leo Szilard Lectureship Award by the American Physical Society and received the 2005 Axel Axelson Johnson Commemorative Lecture award from the Royal Academy of Engineering Sciences, Stockholm, Sweden; the 2009 Frank Kreith Energy Award from the American Society of Mechanical Engineers; and the 2010 Leadership in the Environment Award from Keystone Center. Socolow earned a B.A. in 1959 and Ph.D. in theoretical high energy physics in 1964 from Harvard University.

Dr. Susan Solomon (NAS) is a Senior Scientist at the National Oceanic and Atmospheric Administration in Boulder, Colorado. She made some of the first measurements in the Antarctic that showed that chlorofluorocarbons were responsible for the stratospheric ozone hole, and she pioneered the theoretical understanding of the surface chemistry that causes it. In March 2000, she received the National Medal of Science, the United States' highest scientific honor, for "key insights in explaining the cause of the Antarctic ozone hole." She is also a recipient of the Blue Planet Prize, the Lemaitre prize, the Rossby Medal of the American Meteorological Society and the Bowie Medal of the American Geophysical Union. Her current research focuses on chemistry-climate coupling, and she served as co-chair of Working Group I of the Intergovernmental Panel on Climate Change, which seeks to provide scientific information to the United Nations Framework Convention on Climate Change. Solomon was elected to the National Academy of Sciences in 1992. She is also a foreign associate of the Academie des Sciences in France and the Royal Society of London. She received her Ph.D. degree in chemistry from the University of California at Berkeley in 1981.

Mr. Björn Stigson is visiting professor holding the Assan Gabrielson chair in Applied Corporate Management at the School of Business, Economics and Law at the University of Gothenburg. He has extensive experience in international business. He began his career as a financial analyst with the Swedish Kockums Group. From 1971-82 he held various positions in finance, operations and marketing with ESAB, the international supplier of equipment for welding. In 1983-91 he was President and CEO of the Fläkt Group, a company listed on the Stockholm stock exchange and the world leader in environmental control technology. Following the acquisition of Fläkt by ABB, in 1991 he became Executive Vice President and a member of ABB Asea Brown Boveri's Executive Management Group. In 1995 he was appointed President of the World Business Council for Sustainable Development (WBCSD), a coalition of some

200 leading international corporations. Stigson has served on the board of a variety of international companies and organizations. He is presently a member of the following boards/advisory councils: Prince Albert II of Monaco Foundation; China Council for International Cooperation on Environment and Development; Energy Business Council of the International Energy Agency (IEA); America's Climate Choices Initiative of the US Congress; the Veolia Sustainable Development Advisory Committee and the Siemens Sustainability Advisory Board.

Dr. Thomas J. Wilbanks is a Corporate Research Fellow at the Oak Ridge National Laboratory and leads the Laboratory's Global Change and Developing Country Programs. A past President of the Association of American Geographers, he conducts research on such issues as sustainable development, energy and environmental technology and policy, responses to global climate change, and the role of geographical scale in all of these regards. Wilbanks has won the James R. Anderson Medal of Honor in Applied Geography, has been awarded Honors by the Association of American Geographers, geography's highest honor, was named Distinguished Geography Educator of the year in 1993 by the National Geographic Society, and is a fellow of the American Association for the Advancement of Science (AAAS). Co-edited recent books include *Global Change and Local Places* (2003), *Geographical Dimensions of Terrorism* (2003), and *Bridging Scales and Knowledge Systems: Linking Global Science and Local Knowledge* (2006). Wilbanks is Chair of the National Research Council's Committee on Human Dimensions of Global Change and a member of a number of other NAS/NRC boards and panels. In recent years, he has been Coordinating Lead Author for the IPCC's Fourth Assessment Report, Working Group II, Chapter 7 (Industry, Settlement, and Society), Coordinating Lead Author for the Climate Change Science Program's Synthesis and Assessment Product (SAP) 4.5 (Effects of Climate Change on Energy Production and Use in the United States), and Lead Author for one of three sections (Effects of Global Change on Human Settlements) of SAP 4.6 (Effects of Global Change on Human Health and Welfare and Human Systems). Wilbanks received his B.A. degree in social sciences from Trinity University in 1960 and his M.A. and Ph.D. degrees in geography from Syracuse University in 1967 and 1969.

Dr. Peter Zandan is chairman of EarthSky, a digital media company advocating science as a vital voice in 21st century decision making. He is also senior advisor for Public Strategies, Inc., where he directs strategic initiatives and the research practice group. Peter has helped to launch, lead, and fund numerous business and nonprofit ventures including IntelliQuest Information Group (IQST NASDAQ), the world's fastest growing market research firm in the 1990s; Zilliant, a venture-backed software company; and Evaluation Software Publishing, a K–12 education data analysis software and consulting firm. Peter has also served as a faculty member at the University of Texas at

Austin, where he is a lifetime member of the advisory board of the McCombs Graduate School of Business. He has been selected by Interactive Week as one of the "Unsung Heroes of the Internet" and awarded Ernst & Young's "Entrepreneur of the Year." He also serves on the management committee of the Explorers Club in New York City. He has been active in community organizations including Austin's public television station, St. Stephen's Episcopal School, and Austin's 360 Summit. For his community activities, he has been recognized by the Austin American Statesman as a "Hero of Democracy," by the *Austin Chronicle* as "Best Local Visionary," and by Austin's leading environmental group as "Soul of the City." Peter received his M.B.A. and Ph.D. from the University of Texas at Austin.

Additional Information Regarding the Content of the ACC Panel Reports

Advancing the Science of Climate Change (NRC, 2010a) provides an overview of current scientific understanding of climate change across a range of different areas of interest to decision makers, and recommends steps to advance current understanding. The report focuses on scientific research needed to continue improving understanding of the causes and consequences of climate change as well as the improving and expanding the options available to respond to climate change. It also discusses key attributes and themes for an effective climate change research enterprise, including the research programs, observations, models, human resources, and other activities and tools that are needed. Some report tables of particular relevance include examples of science/research needs related to the following areas:

Table 4.1 Improving fundamental understanding of climate forcings, feedbacks, responses, and thresholds in the earth system

Table 4.2 Human behavior, institutions, and interactions with the climate system

Table 4.3 Vulnerability and adaptation

Table 4.4 Limiting the magnitude of climate change

Table 4.5 Decision support in the context of climate change

Table 4.6 Observations and observing systems

Table 4.7 Improving projections, analyses, and assessments of climate change

Limiting the Magnitude of Future Climate Change (NRC, 2010b) examines how the U.S. can best contribute to global efforts to limit the magnitude of future climate change—primarily through limiting emissions (and enhancing sinks) of GHGs. The report discusses the process of setting goals for U.S. emission reductions; the range of opportunities for limiting emissions from different sources and sectors; the policies needed to assure effective pursuit of "high-leverage" emission reduction opportunities; the resources and policies needed to accelerate technological innovation; the intersection of climate change limiting policies with other issues of major public interest; strategies for integrating federal climate change limiting polices with actions at the local, state, and international levels; and the challenges of developing policies that are both durable over time and flexible enough to be adapted in response to new knowledge. Some report tables of particular relevance include the following.

Table 4.1 Specific policy instruments that can be used (in addition to, or in the absence of, a carbon pricing system) to drive CO_2 emission reductions

Table 3.1 Emission reduction options for non-CO_2 greenhouse gases

Table 5.1 Policy options to influence technology innovation

Table 5.6 Examples of policy impediments to expanding the use of clean energy technologies

Adapting to the Impacts of Climate Change (NRC, 2010c) describes, analyzes, and assesses actions and strategies to reduce vulnerability, increase adaptive capacity, improve resiliency, and promote successful adaptation to climate change. The report discusses the complementary roles of federal adaptation efforts with grassroots-based, bottom-up actions and identifies the key research and information needs for promoting successful adaptation across a variety of sectors and covering a range of temporal and spatial scales. Some report tables of particular relevance include examples of specific options for facilitating adaptation (and identification of entities best poised to implement each option) for the following sectors:

Table 3.2 Ecosystems

Table 3.3 Agriculture and forestry

Table 3.4 Water

Table 3.5 Health

Table 3.6 Transportation

Table 3.7 Energy

Table 3.8 Oceans and coasts

Informing an Effective Response to Climate Change (NRC, 2010d) identifies the range of actors that are making decisions affecting our nation's response to climate change and reviews the different types of decision support tools that are available, or could be developed, to aid those decision makers, including assessments, databases, GHG accounting systems, and "climate services" institutions. It also reviews the different types of decision frameworks that could be used to craft responses to climate change and discusses ways to improve climate change communication through educational systems, the media, and direct engagement with the public. Some report tables of particular relevance include the following.

Table 2.5 Examples of federal departments and agencies that are affected by or involved in decisions about climate change

Table 5.1. Information needs provided by climate services

Table 6.2 Examples of existing GHG emission registries and informing principles

Agenda from the Summit on America's Climate Choices

March 30 - March 31, 2009
The National Academy of Sciences
2101 Constitution Avenue NW, Washington, DC

PROGRAM

March 30, 2009

8:30 AM   WELCOME AND GOALS

Albert Carnesale, Chancellor Emeritus of UCLA & Chair, Committee on America's Climate Choices

Ralph Cicerone, President, National Academy of Sciences

9:15 AM   SESSION 1: WHY IS THIS STUDY NEEDED? PERSPECTIVES FROM STUDY SPONSORS

- **The Honorable Alan Mollohan** (D-WV), Chair, House Appropriations Subcommittee on Commerce, Justice, Science and Related Agencies

- **Jane Lubchenco**, Under Secretary of Commerce for Oceans and Atmosphere and NOAA Administrator

10:00 am   Break

10:30 AM   SESSION 2: KEYNOTE PERSPECTIVES ON CLIMATE CHANGE

- **Robert Socolow, Princeton University**

- **James J. Mulva**, Chairman and CEO, ConocoPhillips

11:30 AM   SESSION 3: WHAT INFORMATION DOES CONGRESS NEED? VIEWS FROM THE HILL

- **The Honorable Bart Gordon** (D-TN) **,** Chair, House Committee on Science and Technology

12:00 PM   Lunch

1:30 PM   SESSION 4: THE CLIMATE CHALLENGE

Moderator: **Diana Liverman**, University of Arizona & University of Oxford

- Certainty and Uncertainty in Climate Science—Framing a Basis for Decisions: **Susan Solomon**, NOAA

- Impacts —The Avoidable and the Unavoidable: **Stephen Schneider**, Stanford University

- Panel discussion: Acting on the certain and the uncertain

  - **Henry Jacoby**, Massachusetts Institute of Technology

  - **Fred Krupp**, Environmental Defense Fund

  - **Charles Holliday**, DuPont

3:30 PM   Break

4:00 PM   SESSION 5: THE AMERICA'S CLIMATE CHOICES STUDY: ARE WE ASKING THE RIGHT QUESTIONS?

Overview of tasks from the Committee and four Panels

- Panel on Limiting the Magnitude of Future Climate Change: **Robert Fri**, Resources for the Future

- Panel on Adapting to the Impacts of Climate Change: **Katharine Jacobs**, Arizona Water Institute

- Panel on Advancing the Science of Climate Change: **Pamela Matson**, Stanford University

- Panel on Informing Effective Decisions and Actions Related to Climate Change: **Peter Raven**, Missouri Botanical Garden

- Committee on America's Climate Choices: **Albert Carnesale**

Panel Chairs respond to questions from the audience.

Question/comment cards will be collected throughout this session.

5:30 PM   Informal Discussion with ACC Members in the Great Hall. *Refreshments provided.*

Tuesday, March 31

8:30 AM   Session 6: Keynote Perspectives on Responding to Climate Change

Introductions: **Albert Carnesale**

- **Mary Nichols**, California Air Resources Board

- **Lorents G. Lorentsen**, Organisation for Economic Co-operation and Development

10:10 AM   Session 7: What Special Challenges Await Us?

Moderator: **William Chameides**, Duke University & Vice Chair, Committee on America's Climate Choices

- Integrating a National Response into a Global Framework: **The Honorable Eileen Claussen**, Pew Center on Global Climate Change

- Vulnerable Ecosystems: **Carter Roberts**, World Wildlife Fund

- Vulnerable Populations & Human Health: **Howard Frumkin**, Centers for Disease Control and Prevention

- Threats to National Security: **R. James Woolsey**, VantagePoint Venture Partners

12:15 PM   Lunch

1:30 PM   Session 8: What Tools Are Available to Meet the Challenges of Climate Change?

Moderator: **Thomas Wilbanks**, Oak Ridge National Laboratory

- Technology Levers: **Robert Socolow**, Princeton University

- Policy and Economic Levers: **Jonathan Wiener**, Duke University

- Regional Impacts & National Assessments: **Jerry Melillo**, Marine Biological Laboratory

- Panel discussion:

  - **Jonathan Schrag**, Regional Greenhouse Gas Initiative, Inc.

  - **Steve Nicholas**, Institute for Sustainable Communities

  - **Heidi Cullen**, Climate Central

4:30 PM   SESSION 9: ARE WE ASKING THE RIGHT QUESTIONS? (TAKE 2)

Moderator: **William Chameides**

- Committee and Panel Chairs and Vice-Chairs take comments and questions from the audience: **Albert Carnesale, Robert Fri, Marilyn Brown, Katharine Jacobs, Thomas Wilbanks, Pamela Matson, Thomas Dietz, Peter Raven, Diana Liverman**

5:15 PM   CLOSING REMARKS: **William Chameides,** Duke University and Vice Chair, Committee on America's Climate Choices

Acronyms and Initialisms

ACC	America's Climate Choices
CDIAC	Carbon Dioxide Information Analysis Center
CDM	Clean Development Mechanism
CMIP3-A	Coupled Model Intercomparison Project Phase 3-A
EPA	U.S. Environmental Protection Agency
EU	European Union
FAO	Food and Agriculture Organization
GATT	General Agreement on Tariffs and Trade
GHG	greenhouse gas
GISS	Goddard Institute for Space Studies
GOESS	Global Earth Observation System of Systems
IPCC	Intergovernmental Panel on Climate Change
NASA	National Aeronautics and Space Administration
NRC	National Research Council
OECD	Organization for Economic Cooperation and Development
RISA	NOAA Regional Integrated Sciences and Assessments
SRES	Special Report on Emissions Scenarios
SRM	solar radiation management
UNDP	United Nations Development Programme
UNEP	United Nations Environment Programme

UNFCCC	United Nations Framework Convention on Climate Change
USAID	U.S. Agency for International Development
USGCRP	U.S. Global Change Research Program
WMO	World Meteorological Organization
WTO	World Trade Organization